The History of the London Water Industry,
1580–1820

THE HISTORY OF THE LONDON WATER INDUSTRY 1580–1820

Leslie Tomory

Johns Hopkins University Press
Baltimore

This book was brought to publication with the generous assistance of the
Johns Hopkins University Press General Humanities Endowment.

Johns Hopkins University Press
2715 North Charles Street
Baltimore, Maryland 21218-4363
www.press.jhu.edu

Library of Congress Cataloging-in-Publication Data

Names: Tomory, Leslie, 1974– author.
Title: The London water industry, 1580-1800 / Leslie Tomory.
Description: Baltimore : Johns Hopkins University Press, 2017. | Includes
bibliographical references and index.
Identifiers: LCCN 2016028540| ISBN 9781421422046 (hardcover : alk.
paper) | ISBN 9781421422053 (electronic) | ISBN 1421422042 (hardcover : alk.
paper) | ISBN 1421422050 (electronic)
Subjects: LCSH: Waterworks—England—London—History. | Water-supply—England—
London—History. | Water utilities—England—London—History.
Classification: LCC HD4465.G7 .T66 2017 | DDC 363.6/10942109033—dc23
LC record available at https://lccn.loc.gov/2016028540

A catalog record for this book is available from the British Library.

Special discounts are available for bulk purchases of this book. For more
information, please contact Special Sales at 410-516-6936 or
specialsales@press.jhu.edu.

Johns Hopkins University Press uses environmentally friendly book materials,
including recycled text paper that is composed of at least 30 percent post-
consumer waste, whenever possible.

To my family

CONTENTS

Many people have helped me in the process of writing this book. Nicola Tynan, Victor Boantza, Bill Luckin, Chris Hamlin, Matthew Hunter, Don Nerbas, and Brian Lewis have provided comments on various parts of the text. Many referees and conference participants have also provided valuable suggestions. I'd also like to thank Victor for his help in obtaining images. Finally, I would like to acknowledge the staff at Johns Hopkins University Press, particularly Elizabeth Demers and Meagan Szekely, as well as Brian MacDonald for his careful copyediting. The research for this book has been supported by grants from the Mellon Foundation and the Social Science and Humanities Research Council, Canada. Parts of this book have appeared in earlier form as articles in *Technology and Culture, Urban History,* and *Social History of Medicine.*

The History of the London Water Industry,
1580–1820

Introduction

Over the course of the nineteenth century, and then more so in the twentieth century, Western cities were slowly reshaped by the construction of large-scale infrastructure that refashioned the daily lives of many of their inhabitants. In 1800, a middle-class Parisian would have walked to work, used a privy or a chamber pot, fetched water from a public pump or well, lit his house with candles, and conducted most communications verbally or by written note. By 1900, his house or apartment would have featured a water closet connected to a sewer system, piped water supplied on demand from a supply network, gaslights fed from a gas utility, and perhaps even electricity. Communications through telegraphs were possible, and telephones were being introduced. He could even have taken a tram or subway to work. Some of this infrastructure, such as the sewers, had become quotidian and commonplace, no longer worthy of special notice, but others, such as the telephone, were in the ascendant, and it would take many decades before they too became part of the infrastructure of everyday urban life. Much remained to be done, certainly, before this infrastructural reforging of the Western city would reach all urban regions. Poorer areas would remain marginalized for some time. But an urban infrastructural momentum had built over the nineteenth century, made possible by changes in engineering and technology, sustained by political systems that favored it, and funded by the accumulating wealth of the industrializing nations of the West. Different cities and nations each had their own particular idiosyncratic character in infrastructure construction, shaped by local social, political, and economic circumstances. But the broad trend toward the network city, interlaced by large infrastructure networks—particularly water, sewerage, gas, electricity, and telephone—was clear.

How did this come to be? Is it possible to give an account for the origins of the infrastructure model that came to be such an important element of Western urban design that is not teleological, seeing the network city as the inevitable outcome of the industrialization and technological development of the West that began in the eighteenth century? Undoubtedly, the network city was made possible by technological and economic changes that gained momentum in Europe through the end of the eighteenth century and that came to be known as the industrial revolution. The expansion of the iron and steel industry, the development of the steam engine, and the rise of electrical technology were some of the changes that became foundations for the infrastructure networks of the nineteenth century. These industrial revolution-era developments enabled the provision of the vast quantities of materials needed to build these networks and provided some of the technology upon which they rested. The industrial revolution, moreover, was a period when infrastructure was constructed on an increasing scale. There were construction booms in the United Kingdom in canals in 1760s and again in the 1790s, and another one in roads in the 1780s.[1] These and other examples of infrastructure construction, such as of ports, meant that by 1800, a precedent had been established for the large nineteenth-century expansion of infrastructure, first new water networks in 1800, followed by gas in the 1820s and then railways in the 1830s and 1840s.[2]

There was, then, movement for infrastructure construction that had been building through the end of the eighteenth century in Britain in particular, and this would become a decisive impulse for the rise of the network city during the nineteenth. Although integrated water supply networks were largely part of the nineteenth-century expansion of urban infrastructure, London's water supply network stands out, both because it was first built long before 1800 and because of its scale and degree of integration. Indeed, integrated urban infrastructure networks were almost exclusively phenomena dating to the nineteenth century. These tightly coupled networks require a greater degree of integration in design and operation because their components, whether technical or human, affect one another relatively intimately and rapidly. In electrical distribution grids in the nineteenth century, for example, engineers had to design generating and load elements to match one another in terms of voltage level and frequencies. Excessive or low voltage could lead to network failure if motors burned out or generators overloaded. Similarly, gas engineers had to closely monitor the chemistry of manufactured gas to ensure that the lighting and heating potential of the gas was adequate.

Furthermore, making such tightly coupled and integrated networks robust necessitates a host of provisions against technical and human deviations from standard or expected behavior. Electrical networks require breakers to limit the damage overloads can cause. Shunt stubs are required to manage the dynamic characteristics of long-distance power transmission by alternating current. In gas networks, pressure regulators and valves ensure smooth and even distribution and help prevent leakages. Within these networks, the human element is just as important as the technological. Workers need to survey the functioning of networks, repair them as needed, and interact with end users. The bills have to be paid for the network to continue running. Finally, the complexity of funding, designing, operating, and maintaining large-scale integrated networks is usually accompanied by a proportionately complex organization, typically a company, but also sometimes a government entity, as in the case of sewers. These organizations reflect the complexity of the physical infrastructure they run and, in turn, develop the infrastructure. The emergence of large corporations has been associated with the emergence of railways, for example.[3]

If this degree of integration increased markedly during the nineteenth century, the few pre-1800 infrastructure networks that existed, such as canals, were by contrast less integrated. They were more tolerant of variations and errors and usually had less need for organizational complexity. As long as the boat was the right size for the canal, the canal remained functional. Catastrophic variations were harder to find. The number of people working for canals was usually small, little more than what was needed to open locks and collect tolls. Boat operators were usually independent actors. Roads were even more tolerant of variations. Very little standardization was required of carriages traveling along them, and having tens of local authorities maintaining sometimes short stretches of adjoining road, while not necessarily efficient, was adequate to keep most of Britain's roads usable, if not comfortable, for the eighteenth century.

London's water supply, however, was different. Originating in the years around 1600 when several water companies were formed, the industry grew from this small base. It was similar in form to that found in other European cities, but by 1700 the industry had attained a notable scale and associated technical and organizational complexity. The London water companies, led by the New River Company, which was larger than all the rest combined, were serving tens of thousands of customers daily, in most cases directly in their houses through tens of miles of pipe. They had dozens of workers maintaining and operating their water distribution networks, fanning out early every morning through the streets to open and shut valves, sending the water to successive areas of the metropolis. Collectors received payments and handled complaints. The rapidly rotting wooden pipes meant constant leaks, and teams of pipe borers, paviours (workers responsible for paving), and laborers to lay and repair pipes. Other workers kept the aqueducts leading to large reservoirs clear of debris or maintained the whirring water-wheels and later steam engines pumping on the Thames.

The builders of London's water infrastructure seemed to have no particular vision for a new kind of network in the early years around 1600. Indeed, London was not exceptional at the time. Most of the technology used was imported from the continent, largely of German origin, and brought by foreign craftsmen. The first water companies struggled for decades to grow, and their scale was not much beyond what could be found in Paris or Hamburg at the time. By 1660, however, the situation began to change. The companies, especially the New River Company, embarked on a rapid expansion, leading to such intense competition from 1685 that profits declined and some bankruptcies resulted. The physical expansion of the infrastructure also brought significant problems in meeting demand, putting the existing infrastructure under stress that it had not been designed to meet. As engineers realized at the time, the problems were not simply of water volume available from sources. The entire network, from source to final distribution, needed to be redone to be better able to meet the demand found in London. These water engineers came up with a series of proposals on how to solve the distribution problems, proposals that would be implemented over the course of the eighteenth century. That the water companies, rather than stagnating, eventually began a new period of expansion shows to what extent these recommendations allowed the companies to reach a new level of integration and scale with their infrastructure. The New River Company took on ever more customers as London's rapid growth continued apace. The London Bridge Waterworks rebuilt its waterwheels repeatedly to maintain and increase its pumping capacity.

By 1820, the end point of this study, the London water industry had existed for more than two hundred years. Although it had its share of critics, regarding both water quality and reliability of supply, it had succeeded in providing about half the buildings of the metropolis with water for a long time. More importantly, it had created a model of integrated network infrastructure in an urban context. In contrast to all urban networks that had preceded it in Europe, including the famed aqueducts of Roman antiquity, the London water model had demonstrated that it was possible to supply a vast city with water, not by means of a reduced number of public fountains that people had to visit, but by directly feeding houses and interior courts. Technologically, this entailed a web of mains that ran beneath almost every street of London, controlled by valves and structured to maintain supply as evenly as possible. It also involved a degree of organizational complexity. The companies employed scores of people ("servants") whose daily task was to keep the network running, by turning valves, interacting with customers, finding leaks, dredging mud, or repairing waterwheels. Legally, all the important London water companies were joint-stock companies, meaning that they had many owners, far beyond what the personal partnerships that dominated business in the eighteenth century allowed. This business form allowed for the investing of more capital for the construction and maintenance of the infrastructure, a crucial feature of most infrastructure construction, as well as the legal status useful to ensure the long-term viability of the companies. This was most clearly the case with the incorporated companies, which therefore had legal personality, although not all joint-stock companies were incorporated.

Thus, by 1800 the London water industry had implicitly shown, without anyone setting out to do so, that it was legally, commercially, and technologically possible to run an infrastructure network within the largest city in the world, one that reached almost all areas, if not all people. Even before 1800, other cities, such as Paris and Dublin, had tried to imitate this model. It was after 1800, however, that the urban infrastructure model would be echoed in water networks in other cities, first in Europe and North America. It would also find imitators in other kinds of networks, beginning with gas in London. The builders of the first gas networks explicitly referred to the London water companies as a model of what a successful urban infrastructure network could look like.

How did this technological achievement come about? The history of the construction of London's water infrastructure was a long one, starting in 1580. Although it faced some difficulties, such as early problems in getting customers and disputes over water quality, over the long term it was clearly effective, enduring, and distinctive, providing water to thousands of houses. There was no single cause for the industry's distinctiveness and success. Rather, various factors came together, the most obvious of which was the sheer size of London. As later chapters discuss, London went from being a mid-tier European city in the sixteenth century to its largest around 1750, when it passed Constantinople, and finally became the largest in the world by the nineteenth century. England was among the leaders in European urbanization from 1550 onward, and before 1800 London was preeminent, taking an ever-greater slice of England's (and Britain's) population. London's was also the seat of government at a time when the royal court was less peripatetic. As northern European trade patterns shifted significantly with the decline of Antwerp in the late fifteenth century, London's port was becoming more important. This development was followed in the seventeenth century by the growth of non-European trade, the ongoing shift of Europe's focus toward the north away from the Mediterranean, and finally the expansion of overseas colonization. The further centralization of much of the administration of the English state and of aristocratic social life with the development of the West End of the city also catalyzed further growth. In the eighteenth century, London was dominant domestically and among the leaders internationally in terms of population, wealth, and commerce. All the while, the increasingly productive English agricultural sector was able to sustain the metropolis.[4]

The size of London provided the masses of people needed to sustain the basic demand for water, but this was not sufficient. Sheer size may explain why London diverged from Hamburg, a city with similar water technology in 1600 that did not experience London's explosive growth thereafter, but not why it differed from Paris or Naples, which were larger for much of the period in question. Indeed, Paris had the same basic water-pumping technology installed around 1600 as well, in addition to repeated attempts to found water supply companies, notably after 1750. The willingness of much of London's population to buy water from a company, to be supplied through pipes to their homes, was also essential. Londoners could, after all, have continued to get their water from wells, public fountains, or rivers via tankards, as Parisians chose to do. Although they may have resisted somewhat initially, they took to the new system with enthusiasm, especially after the 1660s. The New River Company experienced one of its most rapid periods of growth between 1660 and 1690. Many new companies entered the market at this point. Londoners never thereafter lost their willingness to pay water companies for piped water.

This change in consumer attitudes coincided with what Jan de Vries has described as the consumer revolution of the later seventeenth century. Originating in Holland and then spreading to England, it featured the willingness of the middle classes to spend more, especially on domestic purchases. Unlike the ostentatious old luxuries that relished expensive paintings and clothes, the taste for new luxuries expressed itself in the willingness to buy a wider range of household items that were not at the top of luxury ladder. Pipes, clocks, furniture, tobacco, coffee, dried fruit, copperware, and beddings featured in the palette of this new breed of avid consumers. Water was evidently not a new consumer good, but choosing to have it piped in represented a decision for domestic convenience and greater quantities. The consumers of company-supplied water no longer relied on their servants or tankard-bearers to bring them supplies by the jug. They could now rely on more water, delivered to their basement by pipe into a reservoir filled several times a week. It was the later seventeenth-century consumer revolution, combined with London's huge mass of relatively well-off merchants, lawyers, clerks, civil administrators, tradespeople, and shop owners, that allowed the water industry to grow explosively in the later seventeenth century.[5] The growing wealth of these people buoyed the consumer revolution, including demand for water. In the hundred years from 1650, England witnessed rising real incomes per head.[6] Moreover, the price of water remained roughly constant in nominal terms for almost the entirety of the period under investigation, while wages were rising and inflation was present. This meant that, while in 1660 a water connection at about 10 percent of the average yearly wage of a laborer was out of range for most Londoners, by 1800 it had dropped to less than 4 percent. The broadening of the consumer revolution to poorer people, which created demand for the mass-produced goods of the industrial revolution, also meant that ever more of London's population had piped water. By 1800, around 75 percent of houses in the metropolis had water.

Consumption was, therefore, an indispensable part of the reason for the London water industry's long-term growth, but technological innovation was also essential. Throughout this entire history, there were periods of innovation that, taken together, enabled the industry to become established, survive, and expand. The first technology came out of Germany in the years around 1600, notably water pumps and pipe-boring machines. Further innovations around 1700 featured more powerful waterwheels and pumps and especially new thinking about how to build and stabilize the expanding network. This issue was crucial, as the New River Company experienced severe problems in maintaining supply to its rapidly expanding customer base after 1685; although it had enough water entering its reservoirs, it was not able to supply it adequately to its increasingly angry customers. Only after it received recommendations from Sir Christopher Wren and John Lowthorp on how to redesign its network in a systemwide way did it regain its footing around 1720. By deploying more valves, segmenting its network more carefully, and isolating its principal mains from demand disruptions, it managed to fend off its rivals and resume its rapid growth, without encountering the same network problems that had plagued it earlier.

Further innovation was stirring with the introduction of the steam engine for pumping around 1740. The new prime mover had a relatively modest effect at first, but by 1800, after numerous improvements to its efficiency, around 35 percent of water supplied in London was raised by steam engine, increasing to 60 percent by 1820. This advance, combined with the large-scale introduction of iron pipes after 1800, themselves a product of the classical industrial revolution, threw the entire industry into such an upheaval that it underwent one of the most drastic reorganizations of its history. The New River Company lost customers for the first time since the earliest years, and wood pipes, which had been the mainstay of the industry from 1580, were entirely abandoned. The changes also meant that higher-pressure and constantly supplied water was now possible, although it took another eighty years for it to become universal.

The business, legal, and political environment also provided a context within which the London water industry was able to become established and flourish with time. This began with the reliance on for-profit companies that ran the industry during this entire history. Investors and entrepreneurs, initially with significant help from governments, established and funded the companies around 1600. The basic model, once rooted, persisted until the nationalizations in the twentieth century. During this long history, investors stepped forward at various points to create new companies and support existing ones with their capital. This happened especially between 1680 and 1720, and again between 1800 and 1820, although company formation and particularly investment by shareholders occurred outside these two periods. Shareholders in the London Bridge and Chelsea companies, the second and third largest respectively, were willing to invest more capital in the companies to fund expansion in the eighteenth century. The New River Company, by contrast, generated enough capital internally that no new external funds were needed after its first years. These companies and their investors had the incentive to seek and hold onto fee-paying customers, who eventually came to them in such great numbers that, although there were complaints about service and quality at different times, they by and large met demand from all but the poorest people in the city.

The London water industry was founded around a time of innovation and experimentation in legal forms related to commerce, specifically in patents, corporations, and joint-stock companies. All of these had a part to play in the water industry, and all were changing over the long seventeenth century. The issuance of royal patents for commercial purposes, including specific rights and sometimes monopolies, was becoming more common in the late sixteenth century, and different water companies, including the London Bridge Waterworks, were founded through patents. The business corporation was also coming into use at the same time—the early mercantile era—together with the joint-stock form of business organization. All these—patents, business corporations, joint-stock companies—would evolve significantly over the following decades, and the water industry shared in that evolution. The most important among the three was the joint-stock company. It was the business form that the New River Company adopted from the beginning, and which all the large London water companies adopted by around 1700. The large companies created later also took this form. The joint-stock configuration allowed permanent capital and the free transference of shares. It made the pooling of capital from many investors easier than partnerships did, because the latter limited the number of partners to five in most cases and because stock transfers were difficult. With the significant cost of building infrastructure, the joint-stock form helped smooth the industry's development. Although many companies, notably the New River Company, were corporations with legal personality, it is less clear that this feature was as fundamental for the industry's growth in the long run, given that some companies, notably the London Bridge Waterworks, were not established as corporations and yet survived and grew. What the corporate form could provide, if the enabling act granted it, was compulsory access to land for infrastructure construction, something fundamental for all companies. Access, however, could be achieved in other ways, such as through support from local governments, as was the case with the City of London's support for the London Bridge Waterworks.

The power of access granted by governments points to their importance, both local, in form of the City of London, and national, through Parliament and the Crown, in this history. These various entities could make claims, and enforce rights, to water and land, each fundamental to an infrastructure industry whose physical network included long aqueducts, waterwheels on rivers, and pipes running through city streets and fields. Governments proved willing to grant these rights to private companies through royal patents, parliamentary acts, and charters of incorporation, without which the industry could never have functioned. Moreover, and perhaps more importantly, governments did not revoke them once granted over this entire history, with a few exceptions. That this should be so was by no means evident in the shifting multilevel and multipolar political world of seventeenth-century England. The conflict between the Crown and Parliament was intense at times, even exploding into the English Civil Wars. Both Crown and Parliament made possibly conflicting claims to be able to create water companies and grant rights. Parliament was also divided internally, with some blocks representing vested interests that were opposed to water companies, such as landowners who resented Parliament's attempts to bestow access to their land. Although less prone to conflict, the City of London could also be politically unstable, such as when it teetered into bankruptcy at the end of seventeenth century. Investors in water companies had to negotiate uncertainties of a widely shifting political landscape. For this reason, there were very few companies created during the most intense period of turmoil from 1620 until the Restoration of Charles II in 1660, when investment picked up once again with the creation of many new companies. The only arbitrary revocation (as opposed to actions resulting from nonpayment of fees due) of water rights granted to companies by governments happened after the Restoration, when Charles II in 1660 abolished a patent given by the usurper, Oliver Cromwell. With greater political stability after the Restoration, and especially after the Glorious Revolution of 1688, rights granted to water companies were never again seriously threatened.

The Glorious Revolution has featured prominently in the history of England's (and later Britain's) exceptional industrial transformation through the eighteenth century. Douglass North and Barry Weingast have famously argued that following the Glorious Revolution, which displaced James II by William of Orange while shifting effective political control to Parliament, the British state was largely controlled by commercial interests, who ensured that it supported commercial activity, rather than the favor-seeking behavior that dominated governments when monarchs disposed of the state's resources to their rentiers and connections.[7] It was not until the remnants of the Ancien Régime were swept away with the French Revolution that this shift in state interests occurred in other countries. In effect, the Glorious Revolution created institutions, notably secure property rights, that helped mold the behavior of actors such as governments in ways that were conducive to commercial investment and industrial innovation, leading to the industrial revolution of the second half of the eighteenth century. Economic historians such as Dan Bogart have argued the Glorious Revolution did indeed ease investment in infrastructure such as river improvements, roads, and later canals.[8] After the revolution, Parliament met more regularly and was less likely to revoke rights and acts. The Crown was sidelined by the revolution and no longer threatened to usurp rights. To what extent the revolution affected the water industry in the whole country remains open for research. This book examines the London water industry. What is clear is that investment interest in the London water industry was building before the revolution, and although the events of 1688 may have solidified trends in London, the industry was thriving by that time, and many new companies were already forming.

What other factors sustained the industry's growth in the late seventeenth century when it grew rapidly and consolidated? Beside the consumer revolution and London's urban development, the exceptional size and financial success of the New River Company served as a catalyst for further investment in an atmosphere in late seventeenth-century England particularly interested in new projects. It was a time of "improvement," when many people proposed and launched projects of all sorts, aiming particularly for material progress in diverse fields of endeavor, such as agriculture and metalworking. The age of improvement included a boom in patenting; a joint-stock boom with dozens of new companies created; and the financial revolution featuring the rise of the London stock market, the Bank of England, and trading in national debt. As Paul Slack has recently argued, this age of improvement comprised an increasing confidence in the possibility of material progress. The success of some of these new projects established a willingness in the eighteenth century to experiment further, helping to consolidate an "improvement culture" in the country in the eighteenth century.[9] This culture, claims Slack, eventually helped lay the seeds of the "knowledge economy" of the eighteenth century, which Joel Mokyr has argued existed in Britain at that time.[10] For Mokyr, the phenomenon of technological and industrial innovation that came to be known as the industrial revolution was founded fundamentally in Britain's ability to produce, distribute, and use knowledge for practical ends. From the point of the water industry, innovation occurred in spurts, of which the period of the classic industrial revolution was one, featuring the introduction of steam engines and iron. The turn of the eighteenth century, when many issues about network scale were addressed, was another, however. The London water industry was one of the successful industries of the "age of improvement" beginning in the late seventeenth century. It was not founded then, but it did flourish decisively at that time.

This book explores the history of the London water industry in various phases. The first chapter discusses how the industry was established and what Peter Morris, the founding water entrepreneur, brought to London. The second chapter shows how the New River Company was created in a volatile political context. The third and fourth chapters explore the changes in the water industry in the fervid years of the late seventeenth century and demonstrate the important degree of technological innovation sustaining the industry as it grew in scope. The fifth describes the secondary companies, especially the London Bridge Waterworks. The sixth and seventh chapters show how consumers were important and how water purity became an issue for them especially in the later eighteenth century. The eighth chapter discusses how the shock waves of steam engines and iron pipes coming out of the classic industrial revolution struck the industry after 1800.

CHAPTER ONE

The Roots of a New Water Industry

Petruchio: What are they mad? Have we another Bedlam? They do not talk
I hope?
Sophocles: Oh terribly, extreamly fearful, the noise at London-bridge is
nothing near her.

John Fletcher, *The Woman's Prize, or the Tamer Tamed*,
act I, scene 3. (ca. 1605)

Since the late sixteenth century, people walking along London Bridge had
become familiar with the roar coming from the London Bridge Waterworks
perched on the bridge. The wooden wheels slapped the water as they turned
in the Thames's current under the first two arches at the north end of the
medieval stone bridge. Louder than this was the din of the wooden gears
grinding against each other, driving piston heads up and down, and forcing
water into pipes under Fish Hill Street in the city to the north. The waterworks buildings straddled the northernmost piers of the bridge, and it was
from these structures that the builder of the London Bridge Waterworks, a
German (or perhaps Dutch) entrepreneur, had run his new company from
around 1582 up to his death in late 1588. It is not known what his original
German name was, and many variations existed for its anglicized form, but
he is most commonly called Peter Morris because his descendants settled
on that spelling. Morris had come to England before 1572, the year in which
the mayor of Chester hired him to build an aqueduct for the city. Morris's
presence in the country was likely due to his knowledge of water technology. Some German-speaking lands, especially mining areas such as the Harz
Mountains east of Hanover, had been at the leading edge of new water technology, particularly in regard to the pumping needed to keep the miners delv-
ing deeper into the earth in search of wealth.[1] Morris had acquired his skills
somewhere in Germany, and he and some compatriots spread throughout
Europe in the late sixteenth century, bringing their technology with them.
Elizabethan England proved to be a keen recipient for such foreign skilled

workers. Morris was but one of a wave of experienced workers coming to England from abroad in the sixteenth century to undertake many new projects, including those specializing in water supply.

When Morris died in 1588, he could have had little idea that his fledgling enterprise would form the beginnings of a new water supply industry that would, after a century had passed, be serving water to tens of thousands of people in their houses within a vastly expanded city. To be sure, he was not the sole originator of this new industry, nor was he the only innovator behind most of the technology that underpinned it. However, his undertaking introduced two new features—one technical, the other commercial—to the way water was supplied in the city: his water distribution system passed through an extensive physical network of pipes first with hundreds and then with thousands and tens of thousands of branches connecting directly to buildings; and his model was run as a for-profit business. Both of these facets would become important characteristics in how an increasing portion of water was provisioned in London, and both would be defining traits of the industry, at least until its nationalization in the early twentieth century.

The network system, the first characteristic, was based on three components: it relied on a greater supply of water being made available for distribution, such as by means of pumping mechanisms; the pipe distribution network had to be built and maintained, functioning dependably enough to convince users to trust it; and the system connected directly to houses. Technology was one of the pillars in the establishment of the new water industry. Much of the basic technology, such as leaden and wooden pipes, carried over from the earlier period, but it was now deployed on a larger scale. This expansion in scale was made possible in part because of new pumping technologies imported from the continent. Waterwheels, which had existed since antiquity, had come into increasing use for pumping, particularly in Germany with the invention of piston pumps, where they were used in draining mines as well as in urban water supplies. Hamburg, for example, had recently erected such a waterwheel, which was operated by a users' cooperative.[2] The new technology that Peter Morris brought with him acted in effect as a catalyst for the creation of the commercial water industry. Once this new commercial precedent had been set, other people followed Morris's lead and, over the succeeding decades, tried to establish their own water companies. Only some of these relied on new pumping technology. The largest, the New River Company, was supplied by an aqueduct, built at enormous cost, drawing water from the north of the city.

The second pillar in the foundation of these new water companies was the commercial element, and its introduction coincided with broader changes in the structure of commercial enterprise in England and Europe more generally. The later sixteenth and early seventeenth centuries featured the rise of corporations in the increasingly mercantile economies of Europe. Trade had of course been a perennial activity, but with the expansion of trade within Europe and beyond from the fifteenth century onward, new forms of business enterprise developed. The most notable of these was the joint-stock mercantile corporation, epitomized by the East India Company and its Dutch counterpart, chartered in 1600 and 1601 respectively. These were but two of the most prominent of the tens of mercantile companies that received charters of incorporation from their governments from the mid-sixteenth century onward. Most of the early chartered companies were engaged in trade, but some, such as the Society for the New Art of Making Copper (chartered 1572), existed for other purposes, in this case transmuting iron into copper. Unsurprisingly, it failed.[3] The water industry was part of this early modern movement toward the use of new forms of business organization. Up until 1580, the commercial element in water provision in London was limited to the water carriers who hauled their tankards from fountains to homes for a small fee. This changed with the new water companies that began to operate after 1580. In some cases, these were legally partnerships, with fewer than ten co-owners, and in other cases incorporated joint-stock companies, with many more shareholders. The New River Company, chartered in 1619, was the most salient example of an incorporated London-based water supply company. Others would follow over the course of the succeeding century.

The mercantile corporation is the best known among the contemporary innovations in business practices, but there were other trends within which the water industry also fit. In England in particular, the period from approximately 1540 to 1630 was a time of "projecting"—that is, the promotion of new business activities. These "projects" were put forward by people to produce agricultural or other goods for sale that had not been previously made within England in large quantities. Frequently, foreign craftsmen and artisans were enticed to move to England. The new goods included mostly lower-quality commonplace items, such as starch, pins, pots, lace, white soap, and vinegar. The motivations to begin these projects were varied and included the desire to rely less on imported goods; to increase business profits; to find employment for the poor (who had been increasing in number since the upheavals of the first half of the sixteenth century); and, finally, to furnish cheaper items within the rapid inflation of the period.[4]

More broadly than corporations and projecting, the emergence of the London water industry also reflected transformations in the early modern economy. Global trade and the chartered companies were an important part of this, because they were the vanguard of the rise of the business corporation. That the citizens and political leaders in London were willing to countenance, and even foster, a new way of organizing the provision of water for the city speaks to a new commercial mentality. With the exception of the paid water bearers, water provision in medieval cities had long been a communitarian and charitable activity. City governments or local institutions, such as monasteries, would arrange for water supply by building aqueducts from local sources into the city. Water was supplied to a few privileged buildings, as well as distributed at public fountains. This infrastructure was maintained by the local governments through taxes or sometimes fees charged to water bearers for access to the fountains. From 1580, however, the City of London encouraged and supported an entirely different model, one that saw businesses supplying water and selling it for profit. This was no communitarian enterprise, and the shift was decisive, albeit slow in becoming pervasive. Although the City of London did not abandon its public water supply until the eighteenth century, the commercial water industry, once established, slowly took over this task to the point of eventually excluding the government as owner or operator until it reentered the water supply business in the nineteenth century.

London

London at the end of the sixteenth century was a city undergoing rapid transformation. After decreasing in the late Middle Ages due to the Black Death, London's population embarked on a period of growth that would last into the twentieth century, taking it from a second-tier European capital to the top ranks. From a population of around 40,000 people in 1500 and 75,000 in 1550, London reached around 200,000 inhabitants in 1600 and 400,000 in 1650. This crowding of the city, sustained by large numbers of immigrants from the rest of the country, led to periodic epidemics, including influenza and the bubonic plague. The fiercest one, which visited the city in 1665, saw its population drop by 80,000. Due to land shortages in the country and the relatively high wages that even relatively poorer people could earn in the city, immigration was undeterred by such catastrophes. The relative wealth in London was such that Charles Davenant in the 1690s thought that its inhabitants were at least twice as rich as their English compatriots.[5] The city's rapid growth was in contrast not only to other English cities but also to the rest of Europe. Cities in the Low Countries, one of the most urbanized areas in Europe, declined in population in the late seventeenth century. The same was true of most other major European cities, including Paris and Naples. By 1700 London was the largest city in Europe, Constantinople excepted, and it even outgrew that great city by 1750. Although stricken by the depopulating epidemics that all cities suffered, London was largely spared the devastations of war that ravaged many of its peers, nor were famines present since English agriculture proved able to feed the metropolis.

London's local government around 1600 was largely in the hands of the City Corporation, a body responsible for the municipal government of the city dating to the Middle Ages. The area within the City's jurisdiction consisted of twenty-six wards mostly within the old medieval walls north of the Thames, but it also spilled outside of this boundary in many places, including to a small piece of land on the south side of the Thames at the foot of London Bridge. It was governed by a court of twenty-six aldermen who served for life, as well as the lord mayor chosen by the aldermen. There was a Court of Common Council, which met once a year or so when summoned by the lord mayor. It had about two hundred members elected from the City's wards, and they nominated candidates for the position of aldermen, although it was the aldermen themselves who chose which candidate would join their number. The aldermen met frequently and decided most matters in practice, but the Common Council was sometimes active in economic and tax affairs. In the nineteenth century, the Common Council would acquire ascendancy over the Court of Aldermen. The freemen of the City elected the councillors. Householders paying more than ten pounds in rent and thirty shillings in tax per year qualified as freeman. The City's wards were divided into 240 precincts, while another separate division split the City into 108 parishes, which were responsible for poor relief. The City possessed other powers beyond its formal boundaries, such as the government of the navigation of the Thames and the collection of some taxes.[6]

The areas outside of the City were civil parishes of the county of Middlesex, which included the city of Westminster upriver to the west. Westminster, unlike London, had no charter and was governed by a Court of Burgesses from 1585. It did not acquire the same powers as the City of London, and its local parishes were more independent and important in the administration of the city. The rest of Middlesex was governed by justices of the peace for each division within the county, called a hundred, who met for eight sessions per year.[7] Individual parishes within Middlesex also had vestries (assemblies) that had the authority to levy taxes to maintain roads and provide poor relief. Southwark was located south of the Thames across from the City. It was not incorporated or even centrally organized, being divided into parishes, each one governed by burgesses. It was nevertheless referred to as a "borough."[8] Throughout this book, *London* or the *city* is used to refer to the entire urban agglomeration, regardless of political jurisdiction. The proper noun *City* refers to the political entity and the area under its jurisdiction within the larger metropolis.

The population growth of London pushed it far outside of the old medieval city. From 1560, much of this growth took place in poorer areas east of the City, in part because the ever-growing port infrastructure along the Thames in that area provided employment for many sailors, dockworkers, small merchants, and other workers. The northern suburbs and the region south of the river grew especially between 1560 and 1600, while the region to the west grew in the seventeenth century. In contrast to the poor East End of the metropolis, the West End became the site of notable wealth, encouraged originally by the presence of the Crown and Parliament in Westminster. Many peers and gentry built their London residences in the area. Over time, this demographic division became deeply entrenched, marking the social geography of the city to the present day. By 1630, the population in the suburbs equaled that of the City, whose population even declined after 1650. These patterns of growth were to have various consequences for the development of the water industry.

The growth of London also spurred economic changes. Merchants, who had long relied on the export especially of English wool to the continent, expanded into other wares toward the end of the sixteenth century and drove most foreign merchants out of the city. The expanding commerce included a thriving reexport trade whereby goods brought from elsewhere were sent abroad again, sometimes after finishing. The progress of the colonies, largely deprived of the capacity to produce finished goods, also helped boost trade in the late seventeenth century. From being a relatively marginal player in international trade in the 1550s, by 1700 London was challenging Amsterdam as the most important port in Europe, and perhaps a quarter of its population derived its livelihood from the ports in some way.[9]

By the late seventeenth century, London had developed a financial services industry, providing financing, as well marine, fire, and life insurance. Originally closely associated with trade in the sixteenth and early seventeenth centuries, by 1700 the financial sector had become more distinct, branching into supporting other areas, including government debt. Henry VIII had raised his loans abroad, particularly in Antwerp, because its financiers were wealthier and its market more developed than London's. War caused Antwerp's markets to collapse in the 1570s, by which point Elizabeth I could borrow directly from London financiers. The City Corporation itself became involved in royal loans after 1604, raising money from the livery companies or the inhabitants of the City directly.[10] By the 1640s, the Crown was relying on London wealth to meet its needs. A new group of bankers emerged, typically goldsmiths, who took in deposit and lent to the Crown or others, as well as dealing in bills of exchange and issuing bank notes.

An important element of the London economy was centered on the wealthy elites, from the royal court on down. The court had firmly settled in Westminster, and increasingly nobles and wealthy gentry would spend part of the year in the city away from their country estates, bringing their wealth with them. Their luxurious consumption of everything from food to art and entertainment to domestic goods and houses fostered many trades. This demand was met not only through local trades but also by regional networks throughout the country as food, coal, and other goods poured into the city. Banking networks grew to support this regional trade through aiding cash flows by discounting bills.

Of course, most of the city's residents were not wealthy. The largest proportion of the city's inhabitants worked in the textile industry, mostly finishing materials, such as making dresses and hats, rather than producing the raw cloth, which was done in the countryside. Approximately a quarter of the working population was employed in this way in the early seventeenth century.[11] Other important occupations were in the maritime trade and the port, food provision, shops and taverns, metal and leather working, and domestic service. By the 1690s, one in twelve houses in London was a tavern, pub, coffeehouse, alehouse, or the like.[12] The social geography of London followed its economic one because most people lived where they worked, or at least very close by.

The guilds or livery companies that had exercised strong control over the economic life of the City in the Middle Ages declined in importance in the late sixteenth century as the metropolis grew. They exerted less effective control over their respective trades as more business activity took place in the suburbs outside the City itself, even if many of them legally had control over their trade within ten miles of the City's boundaries. Formal apprenticeships, which had been the normal path to entering a trade, declined in importance, and most people simply entered trade without being subject to any form of guild oversight or control. As their effective regulatory powers waned, the guilds transitioned through the sixteenth and seventeenth centuries from being regulators of their respective trades to largely associations for those trades, concerned with social functions, managing property, or charitable activities.[13]

Although London had no university until the nineteenth century, education was also important. The capital was home to many schools, either privately operated or run by churches, guilds, or other institutions. There were also the Inns of Court, which trained lawyers. Other centers of intellectual life within the city included the College of Physicians, chartered in 1518, in addition to the many natural philosophers who later formed the Royal Society in 1665.

Late Medieval and Early Modern Urban Water Supply

London's water supply before the advent of companies was similar to what could be found in many other European cities. Its inhabitants relied on a variety of sources of water that included wells scattered throughout the city, rivers such as the Fleet and especially the Thames, and public fountains fed by lead aqueducts. People would draw supplies from these sources, either directly for their own use or, if they were better off, by hiring water carriers to bring water to their homes in tankards or by cart. There were so many water carriers in the city that they even formed a guild.[14] Among the city's sources of waters, the public fountains were a recent addition. As cities grew in many areas in Europe in the late Middle Ages, city authorities and sometimes other local institutions, such as monasteries, built aqueducts, usually of lead or stone. These aqueducts were low pressure and flowed with gravity, drawing water from springs, lakes, and rivers in the areas surrounding the city. They brought water to cisterns and public fountains, which were known in London as "conduits," the term referring both to the aqueduct and, more commonly, to the destination fountain.[15] Conduits were first built in London by civic authorities around 1230, although the royal palace at Westminster had had one installed nearer 1170. Initially, water from the civic conduits was available to the general public or to water carriers for free, while those using it for a trade, such as brewing, had to pay. From the late thirteenth century, the city government also appointed keepers of the conduits to collect fees, to oversee them, and to guard against attempts to tap into them.[16] By the fourteenth century, all users had to pay.[17]

Water supplied directly to buildings through pipes connected to these aqueducts or conduits was uncommon; most users drew directly from the fountains. There were, however, some privileged buildings with direct supply, including those whose owners had originally built the aqueducts, such as palaces and monasteries. A few other heavy consumers of water, such as brewers, sometimes paid the city for a direct connection into their buildings. In London, brewers were made to sign special contracts from 1345 because they took so much water from conduits that they were depleting supplies, causing complaints from individuals and water carriers.[18] In addition, some people at times attempted to get direct connections (called *quills* because of their small diameters) to aqueducts, both by appeal to local governments and by attaching surreptitious links.[19] In one case of theft, William Campion was jailed in 1478 for tapping into a conduit pipe near his house.[20] Legitimate requests were sometimes denied, but most typically came from wealthy and well-connected people who could bring pressure to bear on the city government, and for this reason many were granted. The last recorded case of a request for such a quill dates to 1662–64, by which point the water companies were offering direct water supply in much of London, obviating the need to badger the City.[21]

London's first conduit was the Great Conduit, built starting in 1245, around the time that many other European cities were having such fountains installed.[22] The Great Conduit was a lead cistern in Westcheap that was rebuilt a few times over the years (fig. 1.1). It was supplied with water from the Tyburn Brook to the west of the City via a lead pipe about six inches in diameter.[23] As the city grew, several new conduits were built, and old ones, including the Great Conduit, were enlarged. Four new conduits were built between 1500 and 1540 with at least four more subsequently, including one endowed by William Lambe at Snow Hill.[24] In total, there were sixteen by the end of the sixteenth century, and more were added later.[25] Particularly after the dissolution of the monasteries, the City of London or the local civil parishes maintained the conduits, and they were paid for through gifts and bequests, such as William Lambe's, or later by public subscription.[26]

New Water Technology

New water technology made possible the shift toward a network model. The late sixteenth century was a time of growing interest in hydraulic engi-

Figure 1.1. The Great Conduit at Cheapside in 1638. P. de la Serre, *Histoire de l'entree de la reyne mere du roy tres-chrestien dans la Grande-Bretaign* (1639)

neering in Europe, sparked in part by the reconstruction of some of Rome's ancient aqueducts.[27] Evidently, not all the technologies that began the slow transformation of London's water supply, such as wood pipes, were novel, but they were used on a much larger scale. This transformation in scale was, however, enabled by changes in production technologies. In the case of wood pipes, the innovation was the pipe-boring machine. Other technologies, such as new forms of waterwheels and pumping mechanisms, were directly involved in water supply. In most cases, the technologies came from other areas in Europe where they had been in use from the late Middle Ages, but many had since undergone some refinements before being introduced to London. Although the technological innovations that gave birth to the new network model of water supply originally came from outside of England, it was in London that they achieved a new scale not found elsewhere.

Waterwheels and pumping mechanism gave access to a quantity of water that surpassed what the conduits, wells, and rivers had been providing. Although waterwheels had been used in the ancient world, they were employed prolifically during the Middle Ages to drive mills. In the later Middle Ages in particular, waterwheels were being put to an ever increasing range of uses, including in slitting mills and fulling mills.[28] In terms of water supply, the wealthy Hanseatic city of Lübeck in northern Germany was recorded as having a bucket wheel providing drinking water in 1294. The wheel supplied fewer than two hundred houses through wooden pipes made from hollowed-out logs. Other German cities similarly introduced bucket pumps with waterwheels to supply drinking water, such as Hanover (1352), Breslau (1386), Bremen (1396), and Bautzen in Saxony (1496).[29] The chain pump was a similar mechanism, using a series of plugs to move water up a pipe (fig. 1.2). Many of these bucket and chain waterwheels, such as the ones in Lübeck and Hanover, were replaced over the course of the sixteenth century with piston-driven pumping mechanisms to provide greater quantities of water. Although it is not clear when piston pumps were invented in Europe, they first appeared in manuscripts in Italy in the fifteenth century.[30] Besides for use in urban water supply, these piston pumps were employed in eastern European mines, as described by Gregorius Agricola (fig. 1.3) in his treatise on mining and metallurgy, *De Re Metallica* (1556). As in urban water supply, the earliest mechanisms used in mining from the late fourteenth century were of the chain-of-bucket type, but by the early sixteenth century, as Agricola attested, piston pumps were finding greater use. These waterwheel-driven piston-pumping mechanisms with their greater capacity spread from Germany and arrived in England at the end of the sixteenth century.[31]

A second hydraulic technology that came to be used much more extensively in London involved wooden pipes. The innovation lay not with the pipe itself but in the scale of their use to connect directly to points of consumption. Wooden pipes, like waterwheels, had a long history going back to antiquity. Although the evidence is not conclusive, they were not much used in Britain during the Middle Ages. Rather, lead and stone were the preferred materials, as they were in France.[32] Wood was much more common in Germany, where it was used for urban water supply (fig. 1.4).[33] One of the earliest cities for which evidence exists with such pipes was Grosal in Saxony, where wooden pipes supplied bronze fountains in the city center around 1200. From the end of the thirteenth century, more cities had such water supply networks. Some inhabitants in Bremen founded a waterworks cooperative in 1396, which, once functioning, served a network of pipes that in

Figure 1.2. A chain pump. Gregorius Agricola, *De Re Metallica* (1561), p. 155

Figure 1.3. A piston pump. Gregorius Agricola, *De Re Metallica* (1561), p. 145

Figure 1.4. Lübeck's waterworks. Note the pipe boring in the foreground. Details from Elias Diebel panoramic woodcut view of Lübeck (1552), reprinted by Emanuel Geibels as *Lübecks Bedrängnis* (1844)

some parts of the city connected to buildings. People paid to join the society, and it cost an additional half mark for a water connection. However, the high cost of these connections meant that this network was used not as a source of drinking water supplied to houses but rather by brewers and bakers for their commercial needs. A few hundred years later, in 1790, the network still had only about 450 connections.[34]

The first German city to have some sort of supply network connecting to houses was built in Hamburg beginning in the sixteenth century.[35] In 1531 and 1535, two piston pump waterworks, similar to those in other German cities, were built on the Alster River, a tributary of the Elbe, while a third mechanism was added in 1620. These waterworks were run as cooperatives whose members were those receiving water, as had been the case with the fountains that preceded the waterworks.[36] The cooperatives provided some direct water connections to buildings through a network of wooden pipes. The cooperatives, which serviced fewer than five hundred connections through thirteen kilometers of pipe into the nineteenth century,[37] were quite small when compared with the later London companies, which had tens of thousands of connections as early as the late seventeenth century. Nevertheless, the nucleus of the network model, with its pumping mechanism and networks of pipes connected to buildings, was created here in Germany.

The third technology that created new possibilities in how water was supplied in the sixteenth century was the pipe-boring machine. Pipe boring had long been done by hand (fig. 1.5), but new machines that could be driven by water or horse power meant that the wooden pipes were produced more rapidly, in increasing quantities, and at lesser expense than had been previously the case. They thereby opened the door to the construction of larger networks of wooden pipes. Better pipe-boring machines, like piston pumps, seemed to have originated in Italy and Germany in the fifteenth century. Mariano Taccola, an Italian polymath and engineer, described such a machine around 1470. By the sixteenth century these machines were being used in new ways. For example, artillery came to be manufactured with bor-

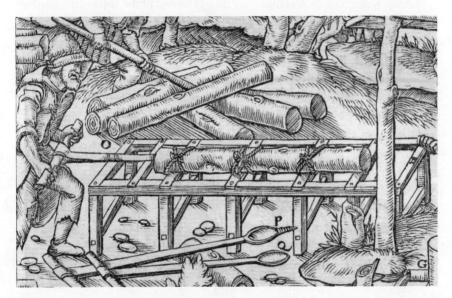

Figure 1.5. A hand-powered pipe-boring machine. Gregorius Agricola, *De Re Metallica* (1561), p. 135

Figure 1.6. Artillery boring mechanisms. Vannoccio Biringucci, *De la pirotechnia* (1540), p. 114

ing mechanisms, as described by Vannoccio Biringucci in *De la pirotechnia* (1540) (fig. 1.6).[38] All these technologies came to England in the sixteenth century.

A Thirsty City

London's water supply was under strain by the end of the sixteenth century. Despite the repeated addition of conduits over the sixteenth century, pressure was building on it because of London's spectacular population growth. This was made manifest by the condition of the city's rivers. Many of the smaller rivers running through the city were becoming quite polluted, effectively functioning as sewers. The City made some efforts to keep them clean, such as in 1502 when the River Fleet was cleansed. The filth returned, however, and although the City in 1589 spent 1,000 marks (£667) to scour the river again, John Stow observed that the "money being therein spent, the effect failed; so that the brooke, by meanes of continuall incroachments upon the banks, getting over the water, and casting of soylage into the streame, is now become worse cloyed than ever it was before."[39] Because smaller rivers were often simply too polluted to be used for any purpose, many were covered over and largely forgotten.[40]

Besides the cleansing of streams, government authorities at both the local and the national level made other efforts to respond to the pressure on existing sources of water. Although the City was preeminent in this effort, Parliament also played a role. It became involved when, after a drought in 1539 and a plague in 1543, it passed the first act related to London's water, probably due to pressure from the City. The new act's preamble described the deterioration in the state of the city's water supply: "Whereof the City of London hath been before this time well furnished, and abundantly served, till now of late, either for faintness of the springs, or for the dryness of the earth, the accustomed courses of the waters coming from the old springs and ancient heads are sore decayed, diminished and abated, and daily more and more be like to disappear and fail to the great discommodity." In response to the dry conditions described in the preamble, the lord mayor had identified fresh sources of water outside the city, such as Hampstead Heath, a hill to the northwest of the City. The new act gave the City the powers to seek out these and other supplies within five miles of the city, even if they were on private land. The act stipulated that the City did not need landowners' permission to access the water (except if it was on royal land) but had to compensate them if water was indeed found and used. In terms of distribution, this act represented a continuation of the conduit model to the extent that it made no specific provision for charging for water, nor did it contain any implication of private supply.[41] The City did not, however, make use of these rights until the 1589 attempt to cleanse the Fleet.[42]

The City's efforts included hiring people with experience with water projects. When it tried to improve its own existing conduit infrastructure, it recruited an Italian engineer named Federico Genebelli (also given as Federigo Giambelli or Gianibelli). Genebelli had lived in the Netherlands and worked in Antwerp in the 1580s on water drainage projects before going to England in 1585.[43] The devastating effect of fireboats of his design on the attacking Spanish fleet during the siege of Antwerp was noted by the English government, which was also engaged in ongoing hostilities with Spain.[44] Genebelli worked in England from 1585 to 1602, and during this time the City repeatedly consulted him on how to improve its water infrastructure.[45] From 1590 to 1592 he was asked to increase the water flow from sources near Paddington to the Tyburn conduit heads, which fed various City conduits. Later, in 1595, he was asked for his opinion about getting water from the Thames.[46]

Because these attempts were not adequate to meet demand, at the end of the sixteenth century the City encouraged entrepreneurs to establish their own water supply enterprises, largely independent of the existing conduit infrastructure. Through their efforts, the for-profit commercial model and network infrastructure were both introduced, and as a result the traditional system of water supply that had prevailed in the city since the late Middle

Ages began to change. While these changes were to have important consequences over the long term, there was also a significant degree of continuity,
and the shift was not revolutionary. For example, the profit motive existed
before 1600, as water carriers charged for their services. In addition, the city
had tried to lease out the Great Conduit to a private contractor for ten years
in 1367, but this experiment failed.[47] What was new in 1600 was that entrepreneurs became interested in selling water to the city using their own infrastructure and operating on an entirely different scale from that of the water
carriers' tankards and carts. The conduit model of water supply supported
largely by the community or charity supplemented by small fees slowly gave
way to one that came to be directed by for-profit private entities.

Patents

One of the legal and political devices that helped establish the London water
industry was the granting of royal patents. The first one in the London water
industry was granted in 1578 as part of a campaign, partly with royal backing,
to encourage the proliferation of the new for-profit projects. More patents
were to be granted to London water companies up to the end of the seventeenth century, and they were a commonly used legal mechanism through
which the Crown became involved with the London water industry.

How patents were granted and used shifted during this time, and the
process frequently became a subject of dispute. The legal form of patents
evolved from medieval letters patent, which were public royal decrees granting certain privileges to individuals. They were not associated with new inventions in any way. In the fourteenth century, Edward III had begun issuing
letters patent granting protection to foreign craftsmen coming to England
to train locals in their trades. The first time the Crown issued a letters patent for a manufacturing monopoly was Edward VI's grant in 1552 to Edward Smythe for a twenty-year monopoly on the production of Normandy
glass. The system flourished under Elizabeth, who used patent monopolies
to encourage commercial projects from 1560, granting fifty-five in total by
1603. These patents attached conditions, such as training locals in the trade
in question, meant to encourage the development of native industries. As
before, these patents were not necessarily associated with new inventions,
although court cases in the 1570s emphasized that they should be related to
the establishment of new trades or products.[48] The method of petitioning for
patents was also established by the Clerks of the Signet and Privy Seal Act
in 1535 and changed little up to 1852, although one modification of note was

that patent specifications became a standard part of the petition in the 1730s. The process was convoluted and involved submitting petitions with fees to many different offices and clerks. A petition could take months to be granted, and contacts at the royal court were essential before 1700.[49]

During the 1580s crown finances came under increasing financial strain, in part because the success of new commercial projects promoted through patents led to a decline in import customs revenues, as many new goods were now being produced in the country. As a result, Elizabeth's patent priorities shifted: new patents were granted on condition that some percentage of the patentee's revenues would be paid to the Crown in return; they were usually not given to foreigners and were not necessarily for new trades. As the Crown granted ever more patents, including on already commonplace items such as the production of playing cards and of starch, they came to be resented and controversial. Most of the criticism centered on whether the public benefits were sufficient to justify the private gains being made.[50] This eventually led Parliament to force Elizabeth to rescind some patents in 1601 and to transfer judicial authority over them to common-law courts away from royal ones.

Patents become more fraught under her successor, James I. His finances were frequently strained by court expenses, and he relied ever more on patents and monopolies for funds. Moreover, he worked to a lesser degree than Elizabeth through the Privy Council, trusting more on his personal judgment in governance, and thereby letting himself be influenced by personal contacts and subjecting the patenting process more to personal intrigue.[51] Eventually, Parliament attempted to restrict James I's ability to issue patents only to new inventions by the 1623 Statute of Monopolies. It also placed patents under the jurisdiction of common-law courts. In practice, however, James and his successors mitigated the force of the statute by wrenching legal authority over patents away from common-law courts back to royal courts, such as the Star Chamber. Over the course of the seventeenth century, as the power of the Crown weakened in favor of Parliament, the statute was interpreted and implemented with less scope for royal prerogative, although patents continued to be granted. From the Restoration of Charles II in 1660, the state law officers, being the attorney general and solicitor general, scrutinized patent applications, effectively preventing them from being granted solely at the Crown's discretion. In practice, however, this scrutiny was at most times perfunctory, and patents were simply registered rather than examined. The result was that despite the scrutiny, patents were still the gift of the Crown. Charles II and James II were able to use them as a means to dispense patronage in a reduced form, and many patentees had contacts at court. The immediate benefits flowing back to the Crown, however, were limited to the now standard fee paid of between seventy and one hundred pounds. There also seemed to be no specific economic policy motivating the granting of patents, such as the importation of continental technologies, as had been the case before 1600. The Glorious Revolution of 1688 restricted the scope for royal initiative in patents even further, but even then the Privy Council occasionally adjudicated patent cases into the eighteenth century. All effective authority passed to Parliament and common-law courts in 1753 when a disagreement between the Privy Council and a common-law court ended with the latter prevailing.[52]

Peter Morris and the London Bridge Waterworks

The first of the new breed of water entrepreneurs in London was an immigrant whose name was variously rendered in English as Peter Morris, Morice, or Moritz.[53] He was the most important of the new water entrepreneurs after Hugh Myddelton, the founder of the New River Company explored in the next chapter. Morris had arrived from the continent some time before 1572, when he signed an agreement with the mayor of Chester to erect a conduit there.[54] Not much is known about him, but he was skilled engineer, probably of German background. He was part of the broader migration to England of continental engineers in the sixteenth century that included Genebelli.[55] He began working on water projects under the patronage of Christopher Hatton, a courtier close to Elizabeth I who would become a member of the Privy Council in 1577 and later lord chancellor.[56] Morris brought with him novel pumping technology, and around 1575 he applied for a patent from the Crown for his "engins and instruments, by motion whereof, running streames & springs may be drawen farr higher, then their naturall leville or course." With Hatton's patronage, Morris secured the patent for twenty-one years in 1578, provided he implement it within three.[57] Peter Morris received his patent before the 1580s when Elizabeth's patent policy shifted toward revenue generation. Morris tried to use these engines "never knowen or used before" in the country to drain fens, but finding this too difficult, he transferred the project to other people.[58] Seeking another potentially lucrative use for his pump, he approached the City about installing a mechanism to supply London. Morris had already in 1574 had contact with the City, even before the patent had been granted him. At that time, he had agreed to build a pumping works to draw water from the Thames to cisterns

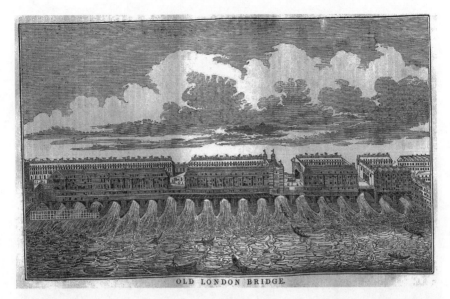

OLD LONDON BRIDGE.

Figure 1.7. A view of London Bridge in 1720. The waterworks are on the left. William Pinnock, *The Guide to Knowledge* (1833) vol. 1, p. 257

in New Fish Street and Leadenhall for one hundred pounds.[59] This first project, however, was slow and dragged on for a couple of years without definite results.[60] Morris blamed the City for this, and he complained to the City and the Privy Council that the City failed to live up to the agreement by not paying the full sum owed him and not making available a piece of land for his work.[61] In 1581, however, he negotiated a much more ambitious agreement.[62] At first, he agreed with the City that he should demonstrate the capacity of his pumps by using them to drive water in lead pipes over the steeple of St. Magnus Church on the north end of London Bridge for which he was paid fifty pounds.[63] After witnessing the demonstration, the aldermen and lord mayor were convinced that his proposed project was viable since "before which time, no such thing was known in England, as this raising of water."[64]

The final deal Morris and the City reached gave him a lease of the first arch of London Bridge, the medieval structure crossing the Thames near the east end of the City (fig. 1.7), for five hundred years at ten shillings per year, where he could build a waterwheel-driven pump mechanism. The bridge had been built between 1176 and 1209 and, as was typical of many medieval bridges, was host to houses and shops along its length. Construction of Morris's waterworks on the bridge began in 1582, and he soon leased on the

Figure 1.8. The Standard in Cornhill in 1599, supplied by the London Bridge Water-works. Robert Wilkinson, *Londina illustrata* (1819), vol. 1, p. 11

same terms the second arch of the bridge from the north shore. The stronger water current there provided more force for driving a waterwheel.[65] Beginning on Christmas 1582, water ran up pipes to houses in Thames Street, New Fish Street, and Grass Street, as well as up to a conduit at the highest point in the City. The conduit was called the Standard and was placed at the east end of Cornhill at Gracechurch Street (fig. 1.8). The City paid for the main to the conduit, and the supply there was initially plentiful.[66]

The water supply from Morris's waterworks was not constant, in part because the Thames up to London Bridge and beyond was affected by tides. Twice a day at low tides, the difference in level between the river upstream and downstream of the bridge was too small to turn the waterwheels, and they lay idle for close to an hour. This feature would be a limitation of the waterworks' supply until it acquired a steam engine in the mid-eighteenth century. In order to be able to serve more customers, the London Bridge Waterworks and the other water companies eventually implemented a system of scheduled supply. Excepting perhaps a very few privileged buildings that had had direct connection to the medieval aqueducts, people had never had an experience of constant water supply. The idea of a constantly available supply of water from the network was very slow to develop and emerged in the nineteenth century only with the introduction of iron pipe networks, as well as high pressure and higher volume pumping.[67] The water companies used workers, called turncocks, to implement a water supply schedule. Every day, these turncocks would fan out throughout the city early in the morning to open cocks throughout the pipe network. The chief turncock would turn valves feeding selected mains according to a weekly schedule. The water would rush from the reservoirs into these mains and down to secondary cisterns and service pipes. The turncocks in each area would open the valves to each of the service pipes drawing water either off the mains or from the cisterns to supply those sections that were scheduled to have water at those times. After a few hours, the valves would be shut down again, including eventually the main valve controlling supply to the entire network. The water supply was cycled between the various mains feeding different areas of the city over various days. This meant that in practice houses received water for two or three hours a few days a week. The system's low pressure could only supply cisterns located in the basement of most homes. The cisterns would fill during the time the water was supplied in the district, and servants would carry the water from the basement to various rooms for use. It is not known exactly when the London Bridge Waterworks segmented its network into different zones and moved to this sort of staged supply, but it and the other early water companies were using it early in their history.

The company was not a complete success, and Morris soon had difficulties in maintaining supplies. Soon after beginning operations, the City wanted to him to provide water to supply the Aldgate ward, which was further away from the Thames along the street from the Standard at Cornhill.[68] The aldermen, however, doubted that Morris's new pump had the capacity to reach the area and agreed with him to clear away houses in Churcheyard Alley near the bridge so that he could expand his works and the "Engyn wh[ich] is thought to be to weake maye be strengethned."[69] Despite this, Morris still had difficulties with increasing the supply of the waterworks. By 1587, Raphael Holinshed reported in his *Chronicles* that its flow to the Standard was "much aslaked," and by 1603 John Stow stated it had run dry.[70]

By supplying public cisterns such as the Standard, the model of water supply that Morris was using was initially similar to the old conduit model. From the beginning, however, he also began to supply individual buildings at a fixed price.[71] The indenture that Morris signed with the City specifically gave him the right to supply water to houses and to lay pipes in the streets so long as he repaired the pavement and gave due notice to the City's chamberlain about the work to be performed. Morris similarly could lay pipes on private land with the agreement of the landowners. Finally, he was allowed to sell water to anyone in the City and to keep any profits.[72] The shift to the network model that Morris initiated was to be gradual, but the increased supply coming from the waterwheels meant that rather than attempting to restrict the number of connections to houses as the City did with the quills tapping into conduits, he was happy to receive the greater income the private connections brought. The supply to the Standard in Cornhill likely ran dry because of increased sales of water directly to houses lower down, the Standard being located at the top of the hill from the bridge. Morris had even complained to the City in 1586 that people were drawing more water into their houses than he wanted.[73]

Although it was Morris who introduced the commercial element and the first hints of the network model to water supply in London, government in form of the City Corporation had a fundamental role to play in bringing about this turn of events. Without the City's support, Morris could not have succeeded, and the City would extend this support to other companies as well. This support began with the legal rights to open the streets to lay pipes, in addition to privileged access to the bridge. Moreover, the City supported Morris financially. It first loaned Morris one hundred pounds in 1580 to aid in the completion of the works, and twice deferred repayment when he was unable to pay.[74] It also ordered the bridge master to "delyver vnto Peter Morrys . . . all suche tooles, instrumts and other necessaryes to worke."[75] The entire project, furthermore, was made possible because the City's common serjeant, Bernard Randolph, gave the Company of Fishmongers "a round sum" to be used to supply the City with water, and the money was passed on to Morris.[76] The City agreed to the gift as it would "profit the whole City, and be no hindrance to the poor water-bearers, who would still have as much work as they were able to perform so far as the water of the conduits would satisfy."[77] Even more financial support came in 1586 when the City judged Morris's engine to be inadequate for supplying new areas.[78] The City lent Morris another one thousand pounds to expand and improve the waterworks.[79] Even this sum was not enough, and in 1587 Morris was lent a further three hundred pounds from an orphans trust controlled by the City.[80] Despite all this support, Morris tired of the project and offered to sell his operations to the City in 1588, but he died shortly thereafter.[81] It was reported in a 1667 lawsuit that the waterworks had cost around twenty thousand pounds to build, although there is no contemporary evidence for this, and the true figure was likely somewhat more than what the City had lent Morris, or around four thousand pounds (see chapter 2 for comparisons).[82] The City in any case declined to purchase the business, and it remained in the hands of Morris's heirs until they sold it in 1700.[83]

Until the company passed out of the family's hands, the company was a partnership among individuals, although its structure became more complicated over time with the addition of a trust controlled by trustees who acted for the Morris family. Originally, the City's agreements were all with Peter Morris personally, but they were written in such a way that he could assign the rights they accorded him to others. He may have taken on partners, and after his death his rights passed on to his widow, Anne Morris.[84] George Digby, who may have married her, then took charge of the company.[85] The company eventually came to be known as the London Bridge Waterworks (LBWW), but never being incorporated, the name was informal.

Other Water Entrepreneurs

The LBWW was not the only water supply business established at this time, nor was it the only one to receive extensive support from the City (fig. 1.9). In an effort to meet the growing demand for water, the City pursued and supported many possibilities for establishing new ways to get more water. One of these was the waterworks at Broken Wharf on the Thames (fig. 1.10), created in 1593 by Bevis Bulmer (1536–1615). Bulmer was a prolific English engineer from Yorkshire,[86] who had worked on various mining projects, including drainage and pumping, for many years. He had grown wealthy working for silver mines near Coombe Martin in Devon from 1587. By 1593, he was ready for a new project and turned his sights to, among other things, water supply in London because the City government lured him there. In that year, the City aldermen wrote him "requestynge him to repayre hither for his helpe in conveyeng of more sweet water to this cytie."[87] Acceding to their request, Bulmer visited the city, and he and the aldermen subsequently continued discussions on ways of getting more water from the Thames.[88] The aldermen eventually agreed in May 1593 that Bulmer himself should build a waterworks.[89] The agreement was that he should pump twenty tuns of water a day to Cheapside.[90] A tun is an antiquated unit of measure used for liquids. One tun is equal to 210 imperial gallons, or about 955 liters, but its definition varied somewhat. His waterworks were soon operating, and in addition to supplying Cheapside, Bulmer also sold water directly to householders via pipes into their houses.[91] The waterworks' pump was driven by four horses rather than a waterwheel, such as used by the LBWW. The pumping station was located at Broken Wharf in Upper Thames Street and supplied Fleet Street.[92] As it did with Peter Morris, the City lent Bulmer money in 1594 to complete his project, beginning with two thousand pounds and followed soon thereafter by another one thousand pounds.[93] The Broken Wharf Waterworks was not, however, a thriving success. Although the City tried to double the amount of water he was pumping to Cheapside, by 1602 Bulmer had defaulted on all but five hundred pounds of the loan.[94] The cost of pumping water by horse was too great to be profitable, as compared to the waterwheels used on London Bridge.[95] In 1604 a citizen of the City, Thomas

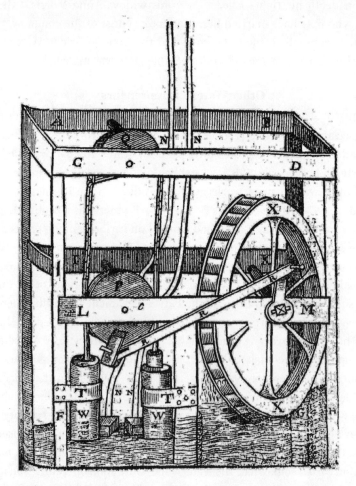

Figure 1.9. The London Bridge Waterworks in 1633. John Bate, *The Mysteries of Nature and Art* (1654), p. 53

Figure 1.10. The Broken Wharf Waterworks (labeled #5 close to Queene hythe). Wenceslaus Hollar, *Long view of London from Bankside* (1647)

Parradine, assumed the debt, and the City transferred the lease of the works to him.[96] Parradine in turn tried to sell half the works to the City in 1608 but with no result.[97] The Broken Wharf Waterworks continued to operate thereafter as a small independent company until purchased by the LBWW in the early eighteenth century.[98]

There were other abortive attempts at establishing water companies that were of little importance except that they demonstrate the range of the City government's willingness to find new supplies of water, and thereby to serve as a catalyst for creating a new water industry. These other new companies also showed that Peter Morris's commercial approach to water supply inspired imitators. In 1593, the same year that Bulmer came to London, the City Council granted Henry Shaw rights to lay pipes through the City streets in order to distribute water from Fogwell Pond on his property in Smithfield to the northwest of the City. The council granted these rights because "the bringing of the sayd water into the cyttie is for the generall good of the same."[99] This enterprise survived only a few years until it was purchased by the New River Company in 1614.[100]

Another City-supported endeavor was Edward Wright's works beginning in 1596. Wright was a mathematician and cartographer and had been a fellow at Caius College in Cambridge from 1585 to about 1594, when he moved to London and became the City's mathematical lecturer.[101] He is best known for his mathematical development of the Mercator projection, published in 1599. Wright, described in the aldermen's minutes as a "masters of arts," was given permission to pump water to Leadenhall in the east end of the City from the Thames.[102] Like the other water businesses, he was given a loan by the City, in his case of £240 in 1600 to complete his pump. He was also given access to St. Botulph's Wharf close to the Tower as a place to build the works.[103] It was the City, moreover, which negotiated with the Muscovy Company to lease more land for the works in 1602.[104] Little further was heard from Wright's waterworks, except that he got into a dispute with George Digby of the LBWW over where they were laying their pipes. The Court of Aldermen ruled that they should keep to the zones they had been supplying before 1602 and not invade each others territory, nor should they "cause to be done any hurt or detryment to any of the pipe or pipes of anie other pa[r]te or partes of the waterwork."[105] Wright's most important role in the water industry was as surveyor for the New River. A last water patent was granted by James I in 1620, this time to Thomas Day for the right to use water arising from springs located in or flowing through Hyde Park, in exchange for an unspecified yearly payment. Day could sell the water in Westminster, and he continued to do so until 1634 when the concession was revoked for nonpayment.[106]

Conclusion

Between 1570 and 1610 the nature of water supply began to shift in a fundamental way in London. Certainly the old ways of supplying water—conduits, water bearers, and wells—continued to be important, in some cases well into the nineteenth century. Nevertheless, some of the new water companies founded in this period endured for two hundred years or more, and constituted an important element of London's urban infrastructure. More importantly, the new companies introduced the profit-driven commercial approach to water supply, coupled with the expansive infrastructure network model. This model combined increased quantities of water, more extensive pipe networks, and distribution to where the water was consumed. New technologies came particularly from Germany. Peter Morris brought them England around 1570 during the age of projects when there was rising interest in improving and establishing new domestic industries by inducing skilled foreign craftsmen to move there, including by granting patents. These newly imported technologies helped bring about the shift to commercial supply selling directly to houses. Although private business was an essential feature of the new model, the City government itself was a key motivator in the birth of the industry. It was seeking many new ways of increasing water supplied to the exploding population, including rebuilding its conduit infrastructure and fostering the new commercial industry. Without the City's efforts to seek out entrepreneurs, and more importantly support them financially and with access to land and streets, they would have floundered. The contemporary trend towards "projects," including importing foreign arts and trades, was also important. The Crown also participated in fostering the water industry in this age of projects by granting a patent to Morris. This was done before patents became a source of revenues for the Crown. The result of all this was that, from no water companies in 1580, London had at least four in 1605.

CHAPTER TWO

The Birth of the New River Company

> For that it is found very convenient and necessary to have a fresh Streame
> of running water to bee brought to the North parts of the of City London,
> from the Springs of Chadwell and Amwell, and other springs in the Coun-
> tie of Hartford not farre distant from the same, which upon view is found
> very fesible, and like to be profitable to many.
>
> 3 James I, c. 18 (1605)

The first twenty-five years of the fledgling water industry saw the London
Bridge Waterworks established and functioning, if not thriving. Changes had
occurred over that period: Morris had died, the company had ceased relying
on the City for financing, and a few other companies had entered the market.
In the first two decades of the seventeenth century, the New River Company
was established, and its advent represented more than just another entry in
the London water supply industry. The New River Company was larger than
the existing companies in a number of ways, starting with the capital required to build its infrastructure. Whereas the LBWW had cost on the order
of £4,000 to build, the New River cost around £19,000 to complete its infrastructure sufficiently to begin operations.[1] By 1620, the total expenses for the
first fifteen years had reached around £32,000.[2] By way of comparison, the
Sovereign of the Seas, one of the navy's largest ships, cost £40,800 to build
in 1634.[3] Dan Bogart has calculated that the completed investment in road
and river infrastructure in the whole country spiked to around £5,000 per
year around 1614 to 1616, with much less before and after.[4] New River's high
cost required around thirty shareholders,[5] many more than other companies,
which, at this point, were partnerships with just a few proprietors.

Because the New River aqueduct ran from sources well north of London
outside Middlesex in the neighboring county of Hertfordshire, the City's political clout was less effective in gaining access to the rights-of-way essential
for such a project. To be able to build across the many lots of private land far
outside London, the builders of the aqueduct had to negotiate with a long list
of landowners, any of whom could pose serious obstacles to its completion,

or even stymie the project entirely. Instead of risking this path, Parliament became involved and granted the project rights of compulsory access to land upon payment of compensation. Resistance from landowners was still fierce, and finally the New River shareholders recruited James I to take an ownership stake in the company directly, helping to overcome the recalcitrant landowners. In contrast to the other companies, which functioned almost exclusively within the City, New River had to consider engagement with political institutions at the national level. As a result, the New River became a corporation to improve its chances of gaining land access rights. It was the first corporation in the history of water supply and was among the earliest group of business corporations, created during the mercantile era that began in the mid-sixteenth century.

Taken together, all these differences between the New River and the other companies meant that its importance for the history of water supply lay not simply with its size. Rather, it acquired distinctive features that it carried forward. However, although its size still set it apart from other companies, it began to lose some of its distinctiveness as the seventeenth century wore on. Just as the New River had done earlier, many new companies acquired patents or charters from the Crown or acts from Parliament in the late seventeenth century, and they too became joint-stock companies. Whereas the LBWW had created the water supply industry by bringing new technology and introducing the commercial element, the New River added the further features of incorporation, the significant involvement of national political institutions, and, most importantly, the joint-stock form.

The New River's status as a joint-stock corporation was particularly suited for its role in infrastructure development, and over the course of the eighteenth and especially the nineteenth centuries, the joint-stock corporation was increasingly used by infrastructure companies. The combination of features brought together by the joint-stock form and incorporation was particularly advantageous. Being a corporation meant having legal stability, which was not tied to individuals; legal personhood, allowing an entity to engage in commercial transactions and legal proceedings in its own name; and perpetual succession, enabling the company to persist regardless of ownership changes. Joint-stock organization allowed the transference of part or all of the ownership without dissolving the company, and it maintained the company's capital permanently, regardless of how shares were traded. Such features in turn eased the pooling of capital. All these traits, but the joint-stock characteristics especially, were important for the scale that most infrastructure projects demanded when the initial investment required was large.[6] The history of the network utilities, such gas, water, and electricity, was dominated by joint-stock corporations in the nineteenth century. The English canals, which boomed at first in the late eighteenth century, were also usually built and operated by joint-stock corporations, the first one dating to 1766. This canal boom had followed on the river improvement projects of the late seventeenth and early eighteenth centuries, which were intended to make watercourses more navigable by widening and deepening them. Some of these were also joint-stock corporations, the first one being the Dun River Improvement Company. It was created by an act of Parliament in 1727 and incorporated in 1733, the first joint-stock corporation in English transport.[7] Dock corporations were also chartered in the eighteenth century, beginning with Hull Dock Company in 1773.[8]

When seen within this context, the use of the joint-stock corporate form in the water industry, particularly with the New River Company, sits at the beginning of a long history of infrastructure construction carried out by joint-stock corporations. By the time that the era of the urban network utilities arrived in the nineteenth century, London's water industry had had more than two hundred years of history operating through joint-stock businesses. The canals had reinforced the usefulness of the model for infrastructure at the end of the eighteenth century, and when new infrastructure networks started to emerge in the nineteenth, the joint-stock corporate form was a well-established choice.

Corporations and Joint-Stock Companies

Patents were legal instruments based on royal prerogative to confer privileges on specific individuals to engage in trade activities. Over the seventeenth and early eighteenth centuries, however, as the scope for this prerogative was reduced, patents were circumscribed to new inventions to be adjudicated by courts of common law, not by the Privy Council. From the Middle Ages, the Crown also had the power to create bodies as corporations, by issuing either a patent or a charter. Incorporation gave these bodies the power to act legally in their own name, as well as perpetual succession. Corporations were used for charitable, social, religious, and other purposes but, with a very few exceptions, not in profit-driven business. Monasteries, universities, towns, and guilds were examples of medieval corporations. Becoming incorporated facilitated how these bodies governed themselves and signed contracts. They could, for example, enter into a contract for a specific service in their own names. Without incorporation, such a contract could be signed only by someone representing the body. Charters usually also granted specific rights and privileges to the corporations, such as to regulate markets within a town. Whereas only the Crown could create a corporation by the sixteenth century, after the late seventeenth century Parliament could compel the Crown to create a corporation by issuing an act creating a company. The incorporating charter was issued by the Crown either when the act gained royal assent or at a later date, as specified in the act.[9]

In the sixteenth century, corporations acquired a new use when for-profit commercial bodies started becoming corporations. Although a very few examples of for-profit commercial corporations can be found the Middle Ages, a fair number were incorporated after 1500, the first being the Merchant Adventurers based in London, which received a charter in 1505. At first, these were regulated companies, meaning that investors contributed financially only to any common infrastructure, such as ships used for trading. Each merchant traded on his own account, and the company itself did not derive profits, nor was there common liability. Such companies were typically awarded special trading privileges, such as a monopoly over the trade with Russia or the Ottoman Empire, and would sometimes negotiate treaties with political entities at target ports. Since participation was personal and limited to single voyages, there were no transferrable shares in these corporations, and there was minimal common governance. To this degree, they resembled guilds as bodies of people engaged in a common trade.

In contrast to regulated corporations, joint-stock corporations featured transferrable shares. Like for-profit corporations, transferrable shares had existed in limited form in the Middle Ages but became more common after 1600 with the incorporation of the Dutch East India Company and its English counterpart, although the Russia Company had been one from 1553 to 1586.[10] Unlike the regulated companies, these new joint-stock corporations made profits on their own account, which they distributed to their shareholders, and they were governed by their shareholders through a governor and directors. Over time, participation in effect came through capital contribution rather than through membership in a trade, as with guilds and regulated companies. This characteristic became stronger with time because the capital of these early joint-stock corporations was not initially permanent. Shareholders reclaimed their invested capital after voyages. The East India Company, for example, did not have permanent capital until 1657, being an ad-hoc joint-stock company from 1600. With permanent capital, its shareholders could no longer reclaim their capital from the corporation. They could do so only by selling their shares to another investor. As with the regulated companies, the Crown also granted a monopoly with the incorporating charter, giving an exclusive right to the company to trade in a specific area or, less frequently, in a specific kind of good. Overseas mercantile trade was the dominant activity of the for-profit corporations: all but three of the pre-1630 English corporations were involved in external trade.[11]

As with patents, corporations evolved significantly over this time. Ron Harris has described the history of the corporation in England before 1800 as having four major stages, and the history of the water industry was also shaped by this evolution. The first early phase ran from the 1550s to the 1620s when the first for-profit corporations were formed. Among the trade companies, the East India Company was the only one to survive in joint-stock form into later periods. Harris argues that the successful use of permanent joint-stock capital by the East India Company into the 1680s made this characteristic attractive for the later trade companies.[12] This same pattern also applies for the New River. Formed in the earliest period of the joint-stock corporation, other water companies sought to emulate its structure in the subsequent period of company formation, beginning especially toward the end of the seventeenth century. The New River was therefore part of an experiment in business organization that succeeded enough to inspire emulation from the late seventeenth century. The company also had its legal idiosyncrasies that were never repeated elsewhere, including a complex share structure split into two blocks.[13]

Another feature of the first business corporations was the frequent, direct involvement of the Crown. James I saw corporations as a way of raising nontaxation revenues. Raising direct taxes required parliamentary approval, and that was not forthcoming, given the strained and increasingly hostile relationship James I and then Charles I had with the body. Because the granting of monopolies through charters with the king as part beneficiary offered another possibility of nontax income, many such grants were given. Most of this activity was directed toward the trading companies, but the New River Company was also a vehicle for the king's desire for revenue. The New River was created in the peak period of this early phase of the business corporation, which ran from 1600 to 1620 and saw forty new incorporations.[14]

The second phase in the evolution of the corporation was one of decline between the 1620s and the 1680s. When Parliament removed James's ability to grant monopolies by patent except for new inventions in 1623, it was also intended to restrict corporate monopolies. From 1625, Charles I continued the struggle with Parliament over monopolies, resulting in chaos and losses for investors, until finally he dissolved Parliament in 1629, which was not to be recalled until the 1640s. As discussed in chapter 1, Charles managed to subvert the restriction on patents. Although the flow of new charters dried up, patents continued to be granted, but not to create corporations. The Civil Wars from 1642 largely put an end to issue until the Restoration in 1660, but even Oliver Cromwell granted patents during the Interregnum. Most of the existing corporations had lost their monopolies by 1660, and the possibility of new taxation with renewed amity between Parliament and the Crown made the income that monopolies had generated less important for Charles II. As a result, few new corporations were created between 1620 and 1688. By the end of this period, however, corporations had acquired a number of important characteristics that they carried forward into the new era: they were joint-stock, with the regulated companies disappearing; their capital became permanent; they were no longer granted monopolies; and the link with the Crown was loosened.

The third stage in the evolution of the English business corporation took place between the 1680s and the 1720s. A joint-stock boom began around 1685, with many new companies forming, reaching around 150 in number by 1695 from a base of around 20 in 1685. Many London water businesses were among these new companies. London's stock market also first emerged around this time, with shares trading daily through stockjobbers and brokers, rather than simply by private sales as they had been the case earlier. The dominant corporations of the period were the moneyed companies (the Bank of England, the South Sea Company, and the East India Company), so called because they became the primary players in holding government debt. This was true despite the latter two originating as trade companies. The moneyed companies have received most interest from historians because they constituted about 85 percent of the value of the stock market.

Not all joint-stock companies created after 1660 were corporations. The unincorporated joint-stock company became an important form of business organization from 1685, especially because charters of incorporation became impossible to get without an act of Parliament after 1720. Investors and entrepreneurs created many companies that functioned as joint-stock companies, with shareholders buying and selling their ownership stakes, without the company ever acquiring legal personality by incorporation. This corporate form had not been created by statutory law. Rather, it had evolved within business practice and common law. Legally, unincorporated joint-stock companies were partnerships that usually functioned through trusts, where a few trustees, who were not necessarily shareholders, legally acted for the business. This arrangement allowed unincorporated companies perpetual succession as the trust endured with changes of trustees. Furthermore, a trust provided a degree of protection against financial liability for the shareholders. Some London water companies were unincorporated joint-stock companies for part or all of their history, the LBWW being the most salient example.[15]

The final stage in the pre-1800 history of the joint-stock corporation began after 1720 when Parliament passed the Bubble Act in an attempt to force more investment into the South Sea Company by making smaller unincorporated companies seem riskier.[16] The act prevented the creation of new joint-stock companies except by act of Parliament, which could have been significant hurdle. The act, however, did not end the formation of new joint-stock corporations. The need for a parliamentary act did not prove to be so significant an obstacle as to prevent new corporations entirely. The canal booms of the 1760s and 1790s saw many canal corporations created by Parliament. In general, businesses that required access to private land by compulsory sale or otherwise usually sought incorporation because they typically acquired the powers they needed in their incorporating acts. Corporations were thus common among canal, dock, harbor, river navigation, and water companies.[17] By the end of the eighteenth century, the joint-stock corporate form was sufficiently familiar that new joint-stock booms took place in the early years of the nineteenth century and included new water companies and the first gas companies.

Entrepreneurs and investors hoping to use patents and corporations had to negotiate the constantly shifting balance of powers between the Crown and Parliament, especially before the Glorious Revolution of 1688. Moreover, political factions within Parliament also changed. As a result, potential investors in the water industry from 1580 onward to the eighteenth century were exposed to substantial political risks. If the Crown granted a patent, Parliament could force its revocation, as had happened under Elizabeth and James I. Furthermore, it was not always clear whether the Privy Council or common-law courts would adjudicate disputes. Finally, Parliament and the Crown both claimed to have certain rights in their gift that were important for water companies. For example, the right of compulsory access to land

and the right to draw water from springs or a river were fundamental to a water company. If one center of power challenged or even overturned rights granted by the other, then it would jeopardize the entire project. In practice, this could have been a problem for London water companies from the industry's founding to around 1700. Although no rights granted to London water companies were ever revoked arbitrarily, with the exception of Cromwell's grant, the companies nevertheless had to lobby either the Privy Council or Parliament at times to gain, preserve, or extend their rights or to fight off rivals. The risks were certainly real, as the experience of other businesses showed. A river improvement act passed by Parliament during the Interregnum in the 1650s granting powers to make improvements on Ouse River navigation was revoked after the Restoration of 1660 because it had never received royal assent, even though the scheme's promoters had already spent twelve thousand pounds on the project. In an earlier case, Charles I revoked a charter granted to Earl of Bedford in 1637 for a drainage project because he supported the parliamentary cause.[18]

Thus, water speculators and investors had to decide which side to approach, with the further complication of the City of London's crucial political role: it too could support or sink a project. The City lobbied Parliament or the Privy Council for or against companies. In the case of Peter Morris, his patron the privy councillor Christopher Hatton had given him the political connections needed to get a royal patent. The City further supported him because it was trying to get water to the growing city. Finally, the revocation of patents forced on Elizabeth by Parliament came too late to affect Morris's patent, which had expired. These different power centers were also important for the New River. When the power balance was definitively settled in favor of Parliament after 1688, the water companies no longer faced such uncertainty. It was finding the votes in Parliament in the face of resistance from landowners or competitors that became crucial.

The London water industry was affected by this long-term evolution of the joint-stock form and the corporation. The LBWW was formed as a partnership but transmuted into an unincorporated joint-stock company around 1700 when that form first came into use. Many London water companies, such as the York Buildings Company, were created during the joint-stock boom from 1685 on. The third of the large London companies, the Chelsea Waterworks, was incorporated by parliamentary act in the final phase of this evolution in 1722. The New River Company emerged in the original flow-

ering of the joint-stock corporation, when the Crown was very actively involved, and when almost all the corporations were mercantile companies.

Myddelton's Politics and the New River Company

The roots of the New River Company began in 1602 when Edmund Colthurst, a former army officer from Bath, decided to add his name to the
small group of entrepreneurs in London who were trying to set up water
supply companies. He planned to cut an aqueduct to bring water to London
from unspecified springs in Hertfordshire, north of the city. His story illus-
trates well the perils investors faced when caught in the duel between the
Crown and Parliament. He obtained a patent from James I for the rights to
water in recompense for his military service in Ireland.[19] The patent granted
him the right to build a water channel six feet across, and he was to provide two-thirds of this water to the City, which planned to use it to cleanse
sewers and ditches.[20] Like the LBWW and other existing companies, he also
planned that the water be "conveyed through pipes and other passages to
particuler howses and places . . . for the necessary uses . . . of persons who
wante water."[21] He started to dig an open channel aqueduct toward the City
but in 1604 became entangled with the City's own attempts to bring water in
an aqueduct, first from Uxbridge from the River Colne or from the River Lea,
and then from the springs of Amwell and Chadwell in Hertfordshire to the
north of the metropolis. These were probably the springs Colthurst had been
planning on using. Likely trying to outflank Colthurst's royal patent, the City
petitioned Parliament for acts giving it rights to water and to build an aqueduct. By this point, Colthurst had spent seven hundred pounds of his own
funds digging about 3 miles (5.3 km) of channel, and wanted to prevent the
City from undermining his project.[22] He first tried to get the City to join his
project and pay for part of the aqueduct and then lobbied Parliament to stop
the grant.[23] Despite his protestations, Parliament in 1606 granted the City's
request for access to water in two acts, crippling Colthurst's project, which
then stalled.[24] The acts allowed the City to build an aqueduct 10 feet (3 m) in
width through the land between the Hertfordshire springs and London. As
difficult as it was for him, the new acts did not entirely destroy Colthurst's
position because the City had agreed in Parliament that it would give some
sort of compensation to Colthurst. Various groups, including Colthurst, lobbied the City to be given the parliamentary rights soon after the original acts
had passed, but these first overtures were rejected.[25] A few years of negotiations followed, with Colthurst exploring the possibility of getting compen-

Figure 2.1. Hugh Myddelton by Cornelis Janssens van Ceulen (1628), National Portrait Gallery, London

sation, as well as of restarting the project by signing an agreement with the City, which would combine the royal and parliamentary rights and remove a potential source of future conflict.[26] The City, however, remained coy, entertaining various proposals.[27] Finally, it decided to transfer its water rights to Hugh Myddelton in 1609 (fig. 2.1). Colthurst himself acquiesced to the plan when he was given shares in the new project, thereby uniting the royal and parliamentary rights.[28]

Hugh Myddelton (b. 1556 to 1560?, d. 1631) was a wealthy goldsmith from an influential Welsh family. In 1576 Myddelton had moved from Denbigh in Wales, where his father had been a member of Parliament, to apprentice as a goldsmith in London. He worked in the City for many years, taking positions such as warden in the Company of Goldsmiths. He was also involved in moneylending as many London goldsmiths were, and by the 1590s had acquired significant wealth. Although living in London, he maintained his connections to Wales by serving as MP for Denbigh for many years, as his father had done. He also became involved in numerous schemes. He acquired a lease for silver, copper, and lead mines in Wales from the Mines Royal Society in 1617 and made a fair profit on the venture, enough to offset the losses he was experiencing in building the New River at the time. Three of his brothers were also London-based merchants, and one of them, Thomas Myddelton, was an alderman and sheriff of the City, sometime MP, and would become the lord mayor in 1616.[29]

For his part, Hugh had been interested in water projects for some time. In 1606 he and his brother Thomas had been on the parliamentary committee considering the City's request for an act giving it powers to get water from the River Colne or Lea.[30] He proved to be shrewd in negotiating many political connections over the first years of the New River's founding. His first success came in 1609, when he managed to displace Colthurst in negotiating with the City over the project after he was given the water rights to the Hertfordshire springs. As with the original agreement with the other water companies, the City provided enormous support to Myddelton and his partners, for he was not alone in the endeavor. The City's support began with the very generous agreement that transferred all the rights given to the City by the acts of Parliament at no charge, leaving any profits in the hands of the company. In exchange, the company only assumed all financial risk. This was much less than had been imposed on Colthurst by his royal patent, which had awarded two-thirds of the water to the City. The parliamentary rights handed over included the power to build the aqueduct through private land on payment of compensation determined by a committee constituted of people representing the areas through which the river passed. Although ownership of the land remained with the original landowner, the act in effect granted the City the power of compulsory access, which it then delegated to Myddelton.[31]

Government-granted rights for the expropriation of or access to private land for the common good was not a legal innovation. It had existed in various forms in medieval law, which allowed land to be expropriated with compensation for the construction of ports, fortifications, or other works.[32] It is not clear that Colthurst had enjoyed any such power of compulsory access to land. He had tried, for example, to negotiate this access with William Cecil

(Viscount Cranborne), and there was no sense that he could compel the cut to be made.[33] The other water companies then operating, such as the LBWW, did not enjoy any parliamentary powers, but they operated almost exclusively within the jurisdiction of the City itself, which could grant powers to lay pipes in the streets. The New River, by contrast, ran through counties outside of the City's control and required something more than what the City could grant.

Beyond these rights, the City also loaned three thousand pounds to Myddelton, after he requested aid in 1614, and further allowed him to use the City's pier to receive pipe at no cost.[34] Not everyone, however, was happy with this generosity shown to Myddelton. Some unidentified parties tried to get the transfer halted in August 1611 because it meant that "that which was intended for a public good shall be converted to a private gain," although without success.[35]

Despite the City's munificence and Myddelton's own wealth, the scale of the undertaking was still beyond his means, and he sought investors in the project, even before he had signed his deal with the City, as he described later in the company's charter of incorporation: "The charge of the said worke greater and heavier than at first was expected the successe thereof doubtfull and the opposicons made against it very stronge besides many other difficulties thought fitt to joyne unto him for helpe therein some other ffriends such as were well affected to the said worke and willing to adventure and joyn in contribucon towards the charge thereof." Colthurst joined the new shareholders when Myddelton agreed with him to absorb the work he had done to date into the new scheme. The two had settled on something even before the deal with the City was finalized.[36] In exchange, Colthurst was awarded a salary, given four shares in the new company out of a total of thirty-six, and was dispensed from having to contribute any further capital for ongoing expenses, as the other shareholders were obliged to do.[37] It is not known how much money Myddelton raised from investors on the original sale, but some shares were sold at one hundred pounds each. He kept thirteen for himself, and by 1619 there were twenty-nine shareholders in all.[38]

The shareholders largely came from close-knit groups; family groupings among them included eight Myddeltons and a connection by marriage, Robert Bateman;[39] three from the Backhouse family, who were related by marriage to two shareholders from the Borlase family; and two from the Hyde family. There were also several political figures among the proprietors, include Henry Montagu, the chief justice of the King's Bench, and a number of MPs: Hugh Myddelton, Robert Killigrew, William Borlase (father and son), Henry Nevill, Nicholas Hyde, Samuel Backhouse, and Robert Bateman. There were also former MPs (Lawrence Hyde, Thomas Myddelton, perhaps Henry Vincent) and many future ones (John Backhouse, John Packer, Marmaduke Rawden). Many of the shareholders also had a commercial background. Besides the Myddeltons, who were miners and bankers, the most experienced was likely Robert Bateman, who was involved in the government of the Skinners' Company. He was very active in the East India Company from its founding, serving as its solicitor, auditor, and treasurer at various points. He was also a member of the Levant Company, the Virginia Company, the French Company, the North West Passage Company, the Merchant Adventurers, and the Massachusetts Bay Company.[40]

As the charter mentioned, the company faced opposition from various parties, most notably landowners who resented having the aqueduct run through their land, as Colthurst had discovered. Recalcitrant landowners lobbied Parliament ferociously against the project, causing work on the New River to halt in 1610. A bill was even introduced to repeal the original acts. Myddelton, however, again showed his political skills. Although the movement to repeal the acts initially received a positive hearing, its progress slowed, partly because the City, as Myddelton's partner, supported his cause, enabling him to prevail.[41] The bill definitively died when James I dissolved Parliament in early 1611, not to be recalled until 1614, by which point construction had finished. After the repeal threat had passed, Myddelton signed a new indenture with the City in 1611, giving him another four years to complete the project. The original agreement had allowed him four years from 1609, but the halt had disrupted his timeline.[42]

Likely motivated by the serious threat to their project that the opposition worked up in Parliament represented, and once again showing political acuity with Parliament not in session, Myddelton and his fellow proprietors recruited James I himself to become involved with the New River Company as an investor in an agreement negotiated in late 1611 and formalized in May 1612. Myddelton and his "adventurers" could have found other investors, but they judged that the king would be a useful ally if the landowners were to renew their opposition.[43] In exchange for half the company in the form of thirty-six new shares, the king agreed to pay half of all costs. He also agreed to allow the company to build on his land at no charge, and to confirm all the rights the City had transferred to the New River. This last point was significant. The king in signing this deal in effect recognized the parliamentary rights now under the New River's control. In an era of intense conflict between the two, Myddelton had neutered the threat of the Crown or Parliament revoking or contesting each other's rights in this case.

In regard to reluctant landowners, the agreement placed the full weight of the king's authority behind the venture. It included a clause stating that the Crown would "withstand and remove all such uniuste and unlawfull impediment which shall or may give lett, disturbance and hinderance to the bringing of the said waters." James also agreed to support any new act concerning the New River put before Parliament. Finally, the agreement established that, although the king had a claim on half of the profits, he was not involved in electing the governors of the company. He kept track of his investment through an auditor he would appoint to verify the company's books regularly, and the company had to report frequently on its progress.[44] This split in ownership into controlling and noncontrolling shares (or moieties) endured long after the connection with the Crown ended in 1630.

After James I became a proprietor, work on the channel progressed rapidly, finishing in September 1613, by which point it had taken four years. Construction had started in 1609, halted in 1610 during the parliamentary debate, and resumed in 1611. The open-channel aqueduct itself was at first around 42 miles (67 km) long, although over the years it was shortened as some lengthy loops were cut out of it with the digging of more direct routes (see fig. 2.2). It was built 10 feet (3 m) wide, as specified in the City 1605 act, and ran with a gentle slope, dropping about 18 feet (5.5 m) over its entire length. In a couple of places, the river crossed over valleys through elevated channels. One, the Highbury frame, was built to cross Hackney Brook. An earthen embankment served as a foundation, with a wooden superstructure carrying the channel erected on top. It was about 420 feet (128 m) long. Another one, the Bush Hill frame of 660-foot (201-m) length, was built to cross the Salmon's Brook near Enfield. The river terminated in Islington at what came to be known as the New River Head (see fig. 2.3). It was located to the north of the city and consisted of a reservoir from which water was distributed to points in the city. Surplus water was also allowed to run into a ditch that led to the River Fleet. Originally, the receiving reservoir was a single round pond, but added capacity came with further reservoirs beginning in 1618.[45] There was also a building at the New River Head called the Water House. The company used it as an office and also controlled flow into the mains leading to the city. The

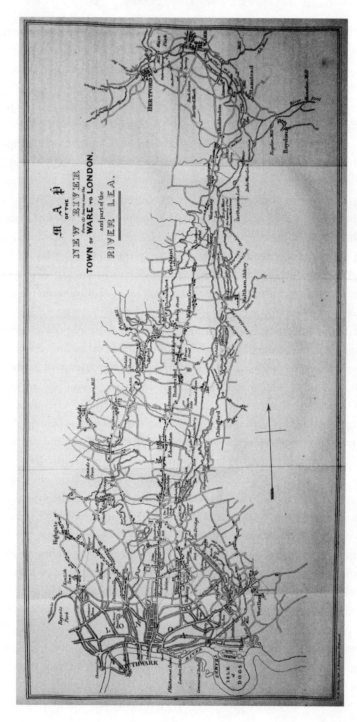

Figure 2.2. Map of the New River. William Matthews, *Hydraulia* (1835), p. 60

Figure 2.3. The New River Head in 1664. Wenceslaus Hollar, *A bend in a river, with houses nearby* (1665). Courtesy Wellcome Trust

two supply springs, Chadwell and Amwell, soon proved to be inadequate, and the company managed to get some additional supply from the River Lea by 1620, although seemingly without specific approval until a royal decree of 1669.[46]

Between 1608 and 1618, Myddelton had proved to be effective in rallying many potentially opposing political forces, and this was most likely his greatest achievement in creating the New River. He had gained the support of the City and the Crown and, indirectly, even Parliament by its absorption of the water and compulsory access rights it had granted the City. Furthermore, Parliament had declined to revoke these rights when landowners brought the issue before it. Colthurst, who had launched the original project, had secured only the Crown's support and had failed when the City and Parliament undermined him. Myddelton made no such mistake, and even if Parliament's support was tepid, he had sufficiently neutralized opposition in it through the City's lobbying, the many MPs among his shareholders, and his own family's long history of service as MPs, so that by 1620 he had prevailed. Myddelton had in effect collected the political support of the Crown, Parliament, and the City and negotiated the shoals of patents, corporations, and acts of

Jacobean England. These machinations would continue, with the New River's incorporation as the next act.

Supplying London

The New River began supplying water to London some twenty-five years after the LBWW but soon grew to be of a comparable size. At first, like the LBWW, the New River fed fountains, cisterns, and basins, in addition to supplying houses.[47] The New River also adopted the method of staging water provision to a different part of its distribution network according to a schedule, rather than feeding it all together. At first, many new customers rushed to sign on. They paid a connection fee, called a fine, which usually amounted to one pound, although it was occasionally more. Within a year of commencing operations, the New River had expanded to such a degree that that George Digby of the LBWW was feeling threatened and complained to the City of "some pre[ju]dice offered to his waterwork by the waterworks of Mr Hugh Middelton."[48] This did not mean, however, that demand for the New River's water was very brisk. In 1614 the company had about 360 customers, and this base saw modest growth, reaching around 765 in 1615 and 1,035 the following year.[49] The company subsequently had difficulties in finding more paying customers (which it referred to as tenants).[50] Growth, however, was slowing, with income from fines paid by new connections almost disappearing around 1618. By that time, five years after the first customers signed on, the king had become disappointed with the venture because it "hath not served to that effect as was expected, either in point of profitt or otherwise." He was considering selling out, or possibly buying the entire company.[51] The number of tenants had grown to about 1,500 that year, but fluctuated around that number for the next twenty years, even decreasing to near 1,350 in 1630 before rising to 2,150 in 1638.[52] The plague year of 1625, when 20 percent of the city's population died, evidently had an effect, and the collector book from around that time is incomplete.[53] This was clearly not the only cause of the stagnation: the fees collected from existing tenants had started to decrease as early as 1620, and recovered only after the plague had passed. Ten years later, the total income had still not reached its 1620 peak, and no dividend had yet been paid. Income from new connections was nowhere near what it had been when the New River first started taking on customers (see fig. 2.4). The easy growth had passed, and the future was more difficult.

The company's mains extended from the easternmost parts of the City of London to its western edge. From there, it had mains running along Fleet

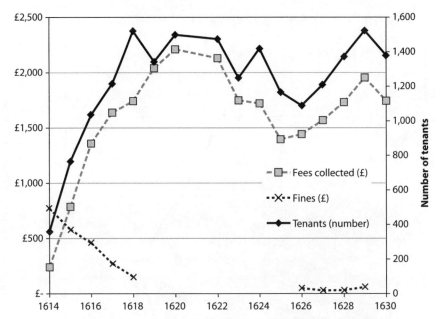

Figure 2.4. Income and tenant count, 1614–1631. Fine indicates one-time fee charged for new connections. Data derived from collectors' books, NA LR 2/25–43

Street and the Strand in the south, and Holborn in the north (fig. 2.5). Its densest zone of service was in the City to area north of St. Paul's. There were also many houses connected along Fleet Street and the Strand. Although pipes were laid in areas to the north of the city walls, customers were relatively few except around West Smithfield. By this point, the mains also overlapped with the LBWW supply zone. Although the New River mostly did not reach the Thames, the two companies were both supplying along Bishopsgate, Cornhill, Leadenhall, and Fenchurch Streets.

Although James did not resolve to change the ownership structure in 1618, he did choose another route that he thought would provide some commercial advantage to the New River, which was to grant it a charter of incorporation in 1619 because it had "not hitherto yeilded such profitt as was hoped for partly by reason of the expences dailie arisinge farr greater and heavier than by the said adventurers was expected and partly for want of power in them to settle the carriage."[54] This new corporation was unique among the charter companies of the period because of the direct ownership stake the king held in the company.[55] In case there were any lingering doubts, the charter once again confirmed the transfer of water rights to the com-

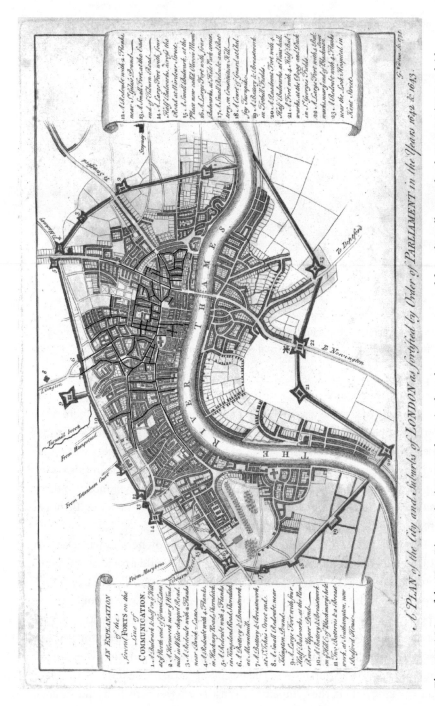

Figure 2.5. Map of the New River mains in 1620. The mains have been reconstructed from the collector's book NA LR 2/37. Overlaid on George Vertue, *A plan of the city and suburbs of London as fortified by Order of Parliament in the years 1642 and 1643* (1738)

pany by the City. It also gave the new corporation, formally entitled "The Governor and Company of the New River brought from Chadwell and Amwell to London," the powers associated with legal personality: the abilities to litigate, to enter into contracts under it own seal, and to enjoy perpetual succession.[56] It went even further by granting the New River a sort of monopoly in the form of a veto right over others supplying the City with water. Other companies could not "attempt or go about to bringe to the said Citties ... anie other river or pipe for conveyinge or bringinge of water from anie place whatsoever without the lycense and Assent of the said [Company]."[57] In practice, however, this proviso had little force because a court case from 1615 restricted the Crown's powers to grant monopolies. The veto's validity was placed in further doubt when the 1623 Statute of Monopolies ended the king's power to grant monopolies, as well as voiding existing ones. Finally, James's own successor, Charles I, would ignore it in 1631 in promoting another London water scheme. The New River thereafter never attempted such a veto over new companies, although its directors briefly mused about the clause in 1666–67.[58]

James's support for the New River extended even to lobbying the City to harry residents to take the New River's water, a prospect that must have worried its competitors but clearly stemmed from the king's interest in seeing it thrive financially. This behavior began before the charter, when the Privy Council, finding in 1616 that "but few, in respect of the generallytie and such for whome it was principally intended, doe take that water," ordered the City aldermen to require that "all such houses within the citty and the liberties, as either out of necessity or conveniencie, may make use the same."[59] In 1617 the Privy Council once again intervened on behalf of the New River, asking the aldermen to delay building a waterworks project at Dowgate so that brewers would be forced to buy New River water instead.[60] Not long before, the Privy Council had intervened to prevent new waterwheels from being installed on London Bridge.[61] In another sign of the closeness between the Crown and the company, in 1622 the New River lobbied the Privy Council to collect unpaid rent and to help prevent vandalism to its property.[62] The king in consequence reminded the lord mayor "to attend to Middleton's water."[63]

There was a downside to the royal embrace. Parliament, locked in an ongoing power struggle with James, was not so enthusiastic about supporting a company that was one of the monarch's projects and proved somewhat hostile on occasion. The original acts had been given to the City, not the company, and Parliament had not agreed to their transfer. When the New River tried to get an act of Parliament to confirm its royal charter, it was rejected, first in 1621, then in 1623, and finally in 1642, after it had reconvened after Charles's long personal rule.[64] Parliament's hostility to James and his granting of monopolies and incorporations meant that most of his plans failed in this area from 1620 onward, when links between the Crown and corporations weakened notably.[65] This distancing from official royal support also happened to the New River. Although James considered buying the entire company once again in 1622, Charles I, who succeeded to throne in 1625, tired of supporting the New River.[66] Like his father James, Charles's interest was fundamentally financial, but the revenues over the years were meager, amounting to around three hundred pounds per year.[67] Finally, Charles sold his half stake back to Myddelton and other investors in 1631 with the proviso that he be paid five hundred pounds per year.[68] This sale came after he had granted a patent to another water supply scheme drawing from Lynchmill Pond that promised him four thousand pounds a year.[69] This scheme failed by 1638, and its promoters joined another group hoping to draw from the River Colne that also failed after they had built 6 miles (9.6 km) of aqueduct,[70] in part due to opposition in Parliament from the New River, now free from royal ties.[71] In 1641 the Colne scheme morphed into yet another led by Edward Ford that also failed in Parliament.[72] From this point until the 1660s, there was little active intervention on the part of the City, Parliament, or the Crown in London's water supply industry, in large measure due to the chaos of the Civil Wars that began in 1642. Myddelton died in 1631 a week after he bought the king's shares. Symbolizing the importance of Myddelton's political connections, and what their termination meant, the City sued his widow for the three-thousand-pound loan granted him during the construction.[73]

Conclusion

The New River Company was created by different forces coming together. The first was the earlier London water companies that had established the possibility of running a business selling water to the city's inhabitants through pipes. These first companies were still small, but the LBWW had survived for more than twenty years by the time Colthurst started his aqueduct, and thirty by the time the New River commenced functioning. This precedent encouraged Myddelton and the other adventurers to risk large sums of their money to build their own project. By 1613, when the water first flowed, the company had expended around eleven thousand pounds, and it would spend seven thousand pounds more in the next year to build its pipes in the city.[74] The new business model Peter Morris had created had been adopted by the most important group to enter the water supply market. There was an important difference, however. The New River's supply was gravity fed, without relying on waterwheels or pumps, a feature whose benefits would become clearer over time; the New River was able to provide more water reliably than the companies subject to the cost, capacity, and reliability constraints imposed by their pumping mechanisms.

The second factor that helped create the New River was the City of London's willingness to foster this new industry as it sought new water supplies. This help was not uniquely given to the New River, as chapter 1 showed, but the help did go beyond what was given to other companies. The delegation of special rights, particularly access to water as a source and to land to construct the channel, which Parliament had accorded the City, was generous and formed the basis of the New River's business model. It would be the legal basis of its model of water supply for its entire history.

The third element fostering the rise of the New River Company was the particular circumstances of early seventeenth-century Crown financing. The king was seeking new revenue streams for himself, and the granting of monopolies to business, particularly through charters of incorporation, was a preferred route. The new external trade companies were the mainstay of this model, but the New River was an experiment in a different line of business. For Myddelton and his investors, this produced many advantages: they received a healthy investment, allowing them to finish the channel; they got the king's assurance of support against ornery landowners; they could rely on the king's pressure to find new customers in the city, even to the point of excluding other competitors; and, finally, they received a charter of incorporation, the second plank in the New River ongoing legal foundation. The end result was that by the 1620s the company was solidly established, and its future depended on finding and meeting sufficient demand for its water.

Finally, Myddelton was a crafty entrepreneur. Unlike Colthurst, he was able to negotiate a potentially treacherous legal and political landscape at a time when government patronage, from either the City of London, Parliament, or the monarch, was essential for the water industry. The early history of the London water industry was deeply intertwined with centers of governmental power, which sometimes conflicted. While some links were simply governments pursuing rent seeking, others were useful for the water industry. Without the City's support for new companies, and the king's and Parliament's willingness to incorporate, grant rights to, and financially and legally support water companies, the industry would not have been able to get underway. Governments had mixed motivations. The City was particularly interested in getting more water to London, while the king wanted income. Parliament, although it had granted the City the water rights in the first place, proved to be hostile to the New River because of its association with the king. But this hostility did not extend beyond refusing to pass an act of incorporation. Myddelton proved adept at handling the political situation, successfully guiding the New River from its foundation, to the commencement of operations, up to his death in 1631.

All this took place during the era of "mercantilism." Mercantilism was at first defined by its critics as the era when governments and mercantile corporations worked together through monopolies and charters to achieve policy goals, such as generating surpluses in trade and controlling colonies, as well as seeking easy unearned income. The term was never used at the time, and its validity and use have been questioned by historians. In recent years, historians of mercantilism have emphasized that rather than being simply a system of collusion between the state and corporations based on specific policies, mercantilism was a complex process. Philip Stern has argued that it was "a product of the private and public pleading, lobbying, treating, and coalition building by companies and their agents to draw the state into service."[75] In this view, the formulation of policy was the result of competition for political and economic advantage on the part of various actors. The history of the formation of the New River Company displays how Myddelton in particular was skilled at influencing governments.

The New River pioneered the use of the corporation and of the joint-stock form, which was possible only with corporations at this time, in water infrastructure. It would, however, be almost fifty years before the next water joint-stock corporation, the York Buildings Company, was created. In the intervening years, some characteristics of the business corporation became more pronounced. Specifically, it was always joint-stock; its capital was permanent; and its association with the government was loosened. For its part, the New River was sui generis as a joint-stock corporation, with its dual class of shares. Nevertheless, by the end of the seventeenth century, its survival made it an attractive model to emulate for the new water entrepreneurs of the period.

CHAPTER THREE

Water in the Age of Revolutions, 1625–1730

Some very happy projects are left to us as a taste of their success; as the
water-houses for supplying of the City of London with water; and since
that, the New-River, both very considerable undertakings, and perfect
projects, adventur'd on the risque of success.

Daniel Defoe, *An Essay upon Projects*, 1697

The period from around 1625 to 1700 was transformative for the London
water industry, as it was for the English economy as a whole. At the beginning of this period, the water industry was still being established. Some companies had been around long enough that they were evidently viable—for
example, the London Bridge Waterworks, with more than forty years of operations. Although much capital and effort had been expended in their construction and upkeep, these companies were no longer getting loans from the
City. Despite their somewhat uncertain profitability over these years, they
managed to stay afloat. Eighty years later, at the end of the period, all these
companies had become quite profitable. The New River Company was so
successful that it undertook an expansion so ambitious that it overreached
and, failing to serve its customers adequately, opened the door for a host of
new competitors to rush in and attempt to grab market share. By that point,
there was no doubting the basic profitability and permanence of the water
supply industry. It had become so successful that the City felt insouciant
about allowing its old conduit infrastructure to decay. The model established
by Morris more than a hundred years before of a network infrastructure run
by companies was firmly rooted.

The consolidation of the London water industry occurred while tremendous transformations were taking place in English political and social life, as
well as in the economic and business environment. The years from 1625 to
1642 were marked by ongoing and increasing tensions at the highest levels
of government, eventually exploding in the Civil Wars, which raged through
various phases. During the first years, London was particularly threatened

by royalist troops, but with the defeat of the royalist cause in 1645 and then the execution of Charles I in 1649, the immediate prospect of violence in the city receded. The Interregnum was ended by the Restoration of Charles II in 1660. The upheavals of this period made for an uncertain economic environment, but as war subsided, economic life resumed its normal course. For the water industry, this phase of its history was one of survival and then slow growth as it was challenged by disease, fire, war, and political upheaval. The largest companies, being the LBWW and the New River, were sufficiently strong that, although there were some serious adverse circumstances, they managed to grow and even to turn a profit for some of this period.

A second phase for the water industry and the broader English economy went from around 1660 to 1685, when greater stability returned to the internal economy at least. The external economy was less stable but manageable. The country fought a series of wars, first with the Dutch, as the two powers came into increasing conflict over overseas trade interests. The colonies, especially in America, were also growing. Although the Restoration in 1660 brought a monarch back to the throne, some of the restrictions won by Parliament over royal power remained in place. Furthermore, an upsurge in entrepreneurial activity manifested itself in enthusiasm for "improving" projects, such as in agriculture and textile production. It was also a period of rapid expansion of the water industry, especially of the New River. The newfound political stability of the era, combined with the swelling wealth of London as it became a leading center of trade, spurred economic growth. The Great Fire of London in 1666 paradoxically reinforced this trend. The fire catalyzed a rebuilding of much of the city, giving an opportunity for the New River to install water connections where none had existed before. New suburbs were constructed to the west of the City, adding huge new markets. The fire also altered the internal dynamics of the water industry because it destroyed the LBWW, allowing the New River to seize an opportunity to invade the LBWW's territory.

The final phase in the seventeenth century ran from 1685 to 1700. Politically, the ascendancy of Parliament over the Crown was definitively consolidated with the Glorious Revolution of 1688, which deposed James II and replaced him with William and Mary. The new settlement circumscribed the monarch's powers, forcing him to ask for funding from Parliament on a yearly basis, thereby guaranteeing that it would sit regularly and creating the conditions where it could regularly check the Crown's activities. The administration of the nation's finances was definitively shifted to Parliament, and the new debt was no longer the Crown's but Parliament's. Furthermore, Parliament gained a decisive upper hand in regulating the national economy over the Privy Council.[1] The financial revolution followed in its wake, with the expansion of London's financial markets in both debt and shares, a new joint-stock boom, and the rise of institutions, notably the Bank of England and the stock market.

These were all to have implications for London's water industry, which was also experiencing some important internal changes. The first was the New River's overexpansion. By 1685, it had taken on far more customers than it could serve, and its erratic supply opened the door for new companies to enter the market. The New River's difficulties were such that it was forced into a prolonged analysis of how it distributed its water, leading to some important technical changes (the subject of chapter 4). The second fresh element present after 1685 was a joint-stock boom The renewed popularity of the joint-stock form after its long quiescence from 1625 led not only to the formation of many new joint-stock corporations but also to the introduction of the unincorporated joint-stock company, which was legally a partnership but effectively behaved as a joint-stock company. This boom ran to about 1700, which was then followed by another up to the 1720s, until it was ended by the inflation and bursting of the South Sea Bubble.[2] Many new water companies were created, leading to fierce competition and depressed profits and dividends for the New River. The LBWW was also affected: in 1703 it passed from the Morris family's hands and was reorganized into an unincorporated joint-stock company. The new owners invested in the waterworks, rebuilding it and expanding it in the process. Taken as a whole, the period from 1660 onward saw the water industry grow rapidly, with the New River taking the lead especially after 1666, but then stumbling from 1685, allowing new entrants into the market whose success put pressure on the New River's profitability.

Slow Growth and Stabilization, 1625–1660

After the founding of the New River, no new water companies were successfully created in London until 1655, although there were numerous attempts. This long dry spell was not the result simply of a lack of interest. There were formidable barriers hampering the aspiration of new entrants. By 1630, the New River and the LBWW had grown significantly, and new companies would have had more difficulty finding a corner of London to supply without entering into direct competition with these two, or even facing active resistance from the now-established players who could lobby to block royal patents or acts of Parliament, as they indeed did. The smaller companies that had been formed in the early years either disappeared or were absorbed by the two dominant players, indicating that large size had brought them a degree of durability. Moreover, the difficulties the New River had in producing profits before 1630 may have discouraged more people from trying their hand.

That the novelty of the business had worn off was shown by the differing opinions about how well the New River was working. On the one hand, the City aldermen, perhaps influenced by Hugh Myddelton's brother Thomas, who was prominent in the City as a former lord mayor and soon to be its MP, expressed satisfaction in 1623 with "the great and extraordinarie benefitt and service this Cittye receiveth by the water brought through the streets of the same by the travaile and industrye of Sir Hugh Middleton."[3] On the other hand, a report prepared in 1622 for James I in his capacity as shareholder described the state of the company in less effusive terms. Its annual expenses were £1,300, while notional income based on contracts was supposed to be £2,844. Arrears were running high, however, and actual collected fees were significantly lower, around £2,130. The author of the report, Sir William Pitt, commissioner of the Royal Navy, did not think that "the workes would be raised to that profitt which they promise to themselves."[4]

The next few years saw little improvement in the profitability of the company, and in an attempt to solve the problem of unpaid fees, the directors resolved in 1625 to cut off the supply of water to those tenants who clearly were better off yet had unpaid debts.[5] Three years later, the situation was even more difficult. Unpaid fees had grown to £4,100, and the directors of the company decided to take a series of steps to try to remedy this. They ordered that a survey of all tenants be made and all people receiving water without a lease be cut off.[6] Some of these connections had even been made surreptitiously, with people secretly tapping into the company's pipes. This sort of theft was such a problem that the directors enlisted the help of local municipal officers to try to suppress them.[7] The City itself was having similar problems with thefts from its conduits and used its officers to search houses where unauthorized connections were suspected.[8]

The New River's struggles with arrears and stolen connections continued, with a notable effect on its finances. Registering income in 1634 of only £2,002 and expenses of £1,601, it was producing worse figures than had been reported to James I in 1623.[9] In 1635 the company redoubled its efforts to cut off anyone with bills more than a year past due, finding some of its own shareholders among the delinquents. It also decided to double the commission it was paying its collectors on moneys received from six pence on the pound to twelve.[10] Some of these measures produced results, and the company finally managed to begin paying dividends in 1633, a practice that would not fail until the twentieth century. It also sought greater revenue by slowly expanding its area of supply and its penetration into the areas it already reached. Originally, its mains reached Smithfield and branched from Fleet Street and Gray's Inn in the west to Coleman Street in the east.[11] Between 1637 and 1639, the directors repeatedly considered the possibility of building mains to Covent Garden to the west of the City but declined to do so because they could still get tenants in the City and its own supplies in the aqueduct were not yet adequate.[12] Covent Garden, which was the Earl of Bedford's estate, was then undergoing urban development with the encouragement of Charles I and the design of Inigo Jones, the surveyor of the King's Works. It was to be first among many grand estate developments in West End that featured the construction of many new houses over the twenty-acre site.[13] Although the possibility of new customers was enticing, the New River directors decided to wait until water at the reservoirs was abundant enough that there was "a contynuall wast at Islington," which had never happened. In addition, the fate of one of the proposed new companies was as yet unresolved. Finally, it was thought best to wait until the "the old worke be brought into better reputacon . . . & the tymes better," presumably reflecting a poor economic environment of the 1630s.[14] How people were using water at this time was hinted at by a patent Charles I issued in 1630 for another water company, in which he noted that "wee are credibly informed that there are very many families both within the Citty of London, and the suburbs thereof and streets adjoining in the county of Middlesex, which want sweet and wholesome water to bake and brew dress their meat and for other necessary uses."[15]

Despite significant barriers to entry, some people were willing to try their hand at creating new water companies. All of these sought royal support, which clearly remained fundamental to the success of a new water company, at least in the eyes of the projectors. This was doubly so because Parliament had been prorogued indefinitely in 1629, not to be recalled until 1640, leaving no alternate route to government patronage at the highest level, unlike the situation the earliest entrepreneurs had faced. The king was indeed keen to act because, as usual, he was seeking funds, and with no possibility of agreement from Parliament, no new funds would be forthcoming from direct taxation. Patents were once again a possibility to be tried, despite the restrictions the Statute of Monopolies had attempted to impose. In the case of water companies, Charles I gave patents to schemes promising to pay him large sums, none of which came to anything in the end. For example, in 1637 he reissued the 1620 patent for the Hyde Park concession, which had been revoked in 1634 for nonpayment.[16] Another scheme that was repeatedly revived in these years was one to bring water from Hoddesdon in Hertfordshire. A first attempt, when Charles granted powers to Sir Nicholas Saunder and others "to convey water by a covered aqueduct from certain springs near Hoddesdon" in exchange for four thousand pounds per year, failed in 1627.[17] Other royal grants for Hoddesdon water were made in 1630 and 1636, but like the earlier proposal they proved abortive.[18] In 1638 the project finally seemed to gain momentum after its various false starts. A commission was established to negotiate with landowners for access to land to build the canal. The promoters raised eighteen thousand pounds by lottery, about half the capital they estimated they needed for the project, but they then ran into financial difficulties after building six miles of the canal. They also faced opposition to an alternate proposal (mentioned in chapter 2) for an aqueduct from the River Colne close to the springs, promoted by Edward Ford.[19]

The backers of these two projects and the New River then fell to attacking one another, with the Hoddesdon scheme promoters writing in a pamphlet that in Covent Garden and surrounding areas there was "a great want of good water . . . especially in the new buildings; the workes already done not being sufficient to supply them, and Middletons water by reason of the foulnesse and muddinesse of it (comming in an open trench) being found by experience not be fit for may uses, and to faile many times for a whole weeke or fortnight together."[20] The charge against the New River was not merely competitive bluster. In 1641 and again in 1644, William Myddelton, Hugh's heir and the company's governor, petitioned the House of Lords on behalf of the company because "many persons have during the late drought made dams and weirs for fishing, and secretly cut sluices in the sides of the river to fill their own ponds, so that but little water has come to the city, to the great increase of the sickness." Earlier, in the absence of Parliament during Charles's personal rule, the company had relied on the City to punish violations of its property, but with Parliament once again in session after a ten years' hiatus, the City referred the matter to it.[21] The Ford and Roberts projects finally merged and promoted a more ambitious plan in Parliament, now including a navigable canal. In retaliation, the New River lobbied to have the

proposal rejected. No more was heard from the proposal with the first stir-
rings of the Civil Wars.[22]

The Civil Wars ran from 1642 to 1651, and although London was spared
most of the military action, it nevertheless created uncertainty in the city
and for the water companies, at least in the early phase when tensions were
particularly high in the city. The water companies once again had to negotiate the now-violent struggle between the Crown and Parliament. Although
the city was staunchly parliamentarian during the first part of the war, many
merchants associated with the incorporated companies supported the royalist cause, presumably because the Crown had been fundamental in establishing these companies.[23] This pattern partly held for New River when a
number of its directors went to fight for Charles's doomed cause, neglecting
the company's operations.[24] However, William Myddelton, Hugh's son and
the company's governor, supported the parliamentary cause, and this helped
protect the company during the war.[25] Parliamentary forces constructed de-
fenses around London consisting of works and forts, one of which was close
to the New River Head.[26] The military action of the war moved away from
London after its early phase, and the little information that remains about
the water companies from this time shows them continuing to expand, especially as a relative calm returned to London. Around 1647, the New River
revived the idea of supplying the Covent Garden area that it had declined to
serve earlier. The directors opened negotiations with the earls of Bedford
and Salisbury who owned much of the land in that area where new suburbs
were being built, and its pipes soon reached St Clement Danes, the civil parish in Covent Garden just to the west of the City along Fleet Street and the
Strand.[27]

In the period between the end of the Civil Wars in 1651 and the Restoration of Charles II to the throne in 1660, there were signs of increased demand
for water that would soon lead to the industry's dramatic growth. In 1654 the
Common Council frequently returned to the question of "the great want of
water in other places within this city now growne very populous." It raised
taxes to improve infrastructure repeatedly, and the aldermen, in order to
preserve as much water as possible, ordered that all quills connected to the
conduits be cut off.[28] This great demand for water stimulated the appearance
of the first new water companies to be formed in forty years. The first came in
1655 when Oliver Cromwell, now occupying the vacant crown's place as head
of state, granted a patent to Edward Ford for a new way of pumping water to

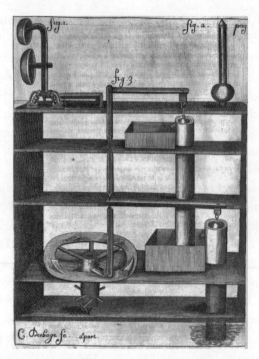

Figure 3.1. Edward Ford's waterworks in 1663. Balthasar de Monconys, *Journal des voyages de Monsieur de Monconys* (1666), vol. 2, p. 29

be used both to drain fens and to supply London, where there was "a great want of water."[29] Cromwell had dismissed Parliament in 1653 and established the Commonwealth under his own rule, once again creating a sole source of official political power for the purposes of gaining water rights. Ford had previously attempted and failed to establish a water and canal company to supply London in 1641[30] but now used his patent to set up a number of small waterworks, including one on the Thames close to the Strand. Called the Somerset House Waterworks, it consisted of a horse-driven piston-pumping mechanism that stored water in a tower.[31] This waterworks, however, soon folded when it fell afoul of the shifting and uncertain politics of the time. It became the only case of a London water company undone by the revocation of water rights. In this case, the newly restored Charles II was not inclined to honor the patents Cromwell had granted. Despite serving "hundreds" of people, Ford's waterworks was destroyed in 1664 after Charles II's mother complained about its water tower obscuring the view from her residence at Somerset House; he wrote to Ford that it "must be removed within three months."[32] Undeterred, Ford and his partners erected another waterworks in Durham Yard off the Strand to the east of Charring Cross, this time sanctioned by the new regime, with a patent from Charles (fig. 3.1).[33]

Growth of the New River, 1660–1700

The second phase of the water industry's growth up to 1700 began in 1660 and was marked by three trends. First, the water companies had acquired a sufficient history of reliability that the City began to lose interest in being directly implicated in water supply, neglecting traditional sources of water and allowing the conduits to decay slowly.[34] Originally, water companies were one component of water supply to the city. The City clearly had not intended to abandon the public conduits once the companies had been established in the early years. Over time, however, the initiative and responsibility for water supply were shifting from public to private hands as the City made ever fewer efforts to expand or even maintain the public conduits. The second trend post-1660 was the rapid rise and then overexpansion of the New River Company. The company had already become the largest long before 1650, but by 1685, it was dominant and continued to acquire thousands of new customers. The company's growth, however, was so rapid that it began to encounter serious problems in supplying all its customers. Finally, the last trend at this time was the emergence of competition. The Durham Yard Waterworks was the first new company, and many more sprang up, especially after the Great Fire in 1666, leading to a period of intense competition after 1685.

The Great Fire of London in 1666 marked a clear turning point in these three trends. The shift from municipal infrastructure toward private enterprise in water supply became definitive after the fire as it reshaped the water supply in London significantly in a couple of ways. First, it destroyed a good deal of London's medieval and early modern water infrastructure in the form of the conduits. Although a few of the City-owned conduits were rebuilt, many were not because the water companies, in a pivotal move, took up the demand.[35] The City, in effect, allowed the slow trend toward the water companies to become definitive by not rebuilding its water infrastructure to its prefire stature. Because the companies had shown themselves to be capable of meeting increasing demand before the fire, the City felt no urgency on this account. As an observer described in 1730, since the Great Fire, "for the conveniency and enlargement of the streets, and likewise by reason the New River Water, contrived by Sir Hugh Middleton, most of these Conduits are taken down, and removed."[36]

The fire also gave a great impulse to the second trend, that of the New River's increasing dominance. Even before the fire, the New River was expanding. In 1656 the directors decided to add a seventh main from the New River Head and to increase the size of its waste (or overflow) pond.[37] These reservoirs were expanded again in 1664 to the detriment of the innkeeper next door, who asked for compensation for the property he had lost.[38] When the fire ravaged much of the City, it left the New River in a good position to take advantage of the reconstruction. The conflagration destroyed the east of the City and those parts close to the river, including London Bridge and the LBWW, while leaving the New River reservoirs and much of its supply areas in the north of the untouched. The New River seized this opportunity by acquiring more land to enlarge its reservoirs in Islington in 1668 and again in 1674.[39] It began to draw more water from the River Lea, leading to a dispute with the river trustees who were concerned about the quantity of water available for the mills located along the river and the depth of the river for navigation purposes. The Crown even ordered the New River to decrease the size of the pipes it was using to draw water from the Lea in 1669, but the dispute simmered over the years, even into the eighteenth century.[40] This new dominance was recognized by contemporaries in a way not observed earlier. For example, William Petty, in considering in 1683 the sevenfold increase in London's population since Elizabeth's reign and the possibility of similar expansion in the future, opined that water could be supplied to meet this need "especially with the help of the New River."[41] Around 1689, a very different source, the satirist Robert Gould, wrote in one of his poems that the New River "through several Pipes supply ev'n half the Town."[42]

The results of the New River's growth became evident in its dividends (fig. 3.2). From around £33 per share in 1640, they reached £64 in 1665, the year before the fire. By 1670 they were at £70, then £145 in 1680, and finally £222 in 1690.[43] Its stock price also followed this trend. Its price had grown so much that by 1695 the New River had the third largest stock market capitalization (£288,000) of any public company in England, after the East India Company (£1,212,000) and the Bank of England (£720,000).[44] Despite this valuation, the New River did not feature prominently in the stock market because the number of its shares was limited by its incorporation to thirty-six adventurers' shares. The royal moiety was also divided into thirty-six

Figure 3.2. New River dividend and share price, 1630–1820. Dividends from LMA ACC /1953/A/258. For share prices, see appendix B.

shares soon after its creation.[45] Half of the company's shares remained undivided, meaning that they traded infrequently at huge values.[46] The number of shareholders continued to be very small well into the eighteenth century. An audit book from 1770 listed eighty-five share holdings, but with duplicates between the king's and adventurers' moieties, there were sixty-nine different shareholders.[47]

Some documents provide information about the first shares. The earliest indicated the sale of two shares to Sir Henry Neville in 1612 for £100 each, with further sums owing should company expenses exceed £6,400. These were to be borne proportionately by all shareholders except Colthurst's four shares, in recognition for his initial expenses. The later practice of initial investors paying only a portion of the nominal share value with the remainder to be paid via calls was not yet current.[48] By 1619 there were twenty-nine shareholders or adventurers.[49] The total amount these shareholders paid in after their initial investment is not known, but on the basis of the expense books provided to the king for his own contributions up to 1630, it was

£18,524, or about £289 paid up value per share.[50] By that point, the company was close to profitable, paying its first dividend in 1633. There were no records of other calls after that date, and the uninterrupted dividends suggest that none were needed, a claim asserted by the company in a report given to Parliament in 1834.[51] The long-term profitability of the company was such that those who held onto their shares would have been making yearly returns, adjusted for inflation, of 20 percent by the 1660s and between 60 and 80 percent from 1690 to 1800. A few shares passed on through families into the eighteenth century, and one share donated by Myddelton to the Goldsmiths' Company was held until the New River was nationalized.[52] A share held from the time of full paid-up initial investment of £289 in 1630 to 1805, when it would have sold for £15,840, would have accumulated £36,031 in dividends for a total return of 103 percent per year in nominal terms or about 75 percent when adjusted for inflation. An investor buying shares later on would not have received anywhere near these returns with the steep price increases the shares enjoyed. In terms of dividend yield on current share price, the New River returned about 8 percent to 1660, and then dropped to 6 percent as the political situation stabilized. The yield dropped once again to 4 percent just before 1700 and stayed around that level throughout the eighteenth century, comparable to the yield on long-term government bonds (fig. 3.3).[53] Investors evidently felt that the company's prospects and its dividends were quite solid.

The New River Company's charter established how it was governed. To be "of the Company," a shareholder had to own an entire share in the adventurers' moiety, the king's moiety being excluded from governance by the original contact creating it. The number of members was further limited to twenty-nine, the original number of shareholders. With the subsequent division of shares, there apparently never was more than that number of whole shares, avoiding a potentially thorny problem. These members elected from their own number the company's governor, deputy governor, and treasurer. The original charter did not use the word *board*, typical in the nineteenth century, or even the term *court of directors*, more common before 1800. It did not even use the word *director* because this group of adventurers constituted the company itself. The charter specified that the governor and other officers were to be "of the said Companie" and to be chosen "from tyme to tyme" and had the power to call "Courts, Assemblies, Councells and Consultacons . . . for the better direccon . . . of the affaires and businesses concerninge the same Companie." The charter contained provisions for the quorums needed

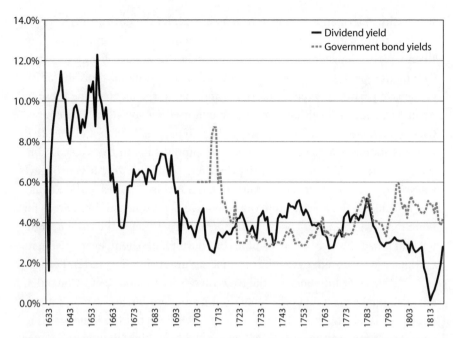

Figure 3.3. Long-term government bond yields and dividend yields on New River stock, 1630–1820

to make certain decisions, such as five for making bylaws, signing a water lease, or paying expenses, or seven to elect the officers. This last item was to be done on a yearly basis.[54] The quorum needed for operational decisions like paying expenses ensured that at least five shareholders were regularly involved. All this meant that there was quite a restricted company of people who had control of and participated in the New River's governance, even within the small number of shareholders. Those outside the controlling group were not even formally represented. They were simply not part of the company, although they had full claim to its profits.

The New River's growth meant that its competition with other water companies grew fiercer. In many cases, this competition was fought not only through price and quantity of water served but also by means of government patronage, including by the novel attempt to lay claim to exclusive zones of supply. In response, the Privy Council generally tried to moderate the competition but rejected any claims to exclusivity. One dispute occurred between the New River and Ford's waterworks in Durham Yard, which by then had passed on from Ford to other owners, including Robert Vyner, who was a City alderman and later lord mayor. In 1666 the new proprietors got into a dispute with the New River when someone tore up some of the New River's pipes, depriving the inhabitants of the New Market area of water. The matter came before the Privy Council, which ordered representatives of the two parties to appear before it, "and that in the meane time there be no disturbance or disorder offered or committed on any side." The council tried to mediate a settlement between the two parties.[55] Vyner and the other proprietors made a claim to an exclusive supply zone in the West End as their "quarters" and tried to get the Crown to force the New River to leave the area (St. Clement Danes, St. Martin in the Fields, St. Giles in the Fields). They made this demand despite the New River's presence in those areas for more than twenty years. The New River counterpetitioned the Crown, arguing that it had the right to lay its pipes by charter and patent, even raising its vestigial power of veto over other companies granted in its charter. Perhaps reflecting greater influence on the New River's part on the Privy Council, it decided to take no action other than to urge moderation and attempt mediation.[56] This left the New River free to operate there, but the issue was rendered moot toward the end of 1667 when the New River bought out Vyner and his partners in 1667 for £6,100.[57] Their works were not large, supplying a few hundred people and making a profit of about £300 per year.[58]

The New River got into a further quarrel with another small water company at this same time, and like the Durham Yard Waterworks, it also turned on political connections. A small waterworks had been established somewhere in Marylebone to the west of the City and was run by William Smith and John Hooke. It likely also originated with Edward Ford, who had been given permission to build a waterworks there at the same time as his Durham Yard Waterworks. In 1668 this dispute came before the Privy Council, with the Marylebone Waterworks claiming that the New River was invading its territory in Covent Garden and "selling as under rates, to the distruction of the said water worke of Sir William Smith." The Privy Council appointed a committee to forge an agreement between the two parties "to moderate and sett the prices at which the parties shall sell their water, that soe reasinable rates being sett they may not undersell each other." The negotiations between the two dragged on, but they finally agreed to a price-fixing arrangement in 1669 that allowed both parties freedom to lay pipes anywhere and to charge a minimum of twenty-one shillings per house of ten rooms or more. They also agreed to charge for water sold from standpipes in the street for a penny for fourteen gallons. There was a residual disagreement about where the New River could set up its standpipes, with the company insisting that by law it could do so anywhere it pleased. On this point, the Privy Council's report to the king merely stated that the New River should generally be favored because it paid the Crown five hundred pounds per year, while the other company paid nothing.[59] In effect, the Crown had moderated the competition between the two and had blessed the New River's growth into another company's area.

Finally, the LBWW got into a clash with the New River in which it tried to claim an exclusive right to supply water to a specific area, to be enforced by government. The LBWW had been around for much longer, and its dispute with the New River was occasioned by the Great Fire. The owners of the LBWW were slow in rebuilding, taking more than two years due to a falling-out among various heirs to Peter Morris and the trustees who cared for the works. Mary Morris, the widow of Peter's grandson John Morris, sued her brother-in-law Thomas to force him to rebuild the works. She claimed that he was intentionally delaying reconstruction in order to lower the annuity she received from the trustees. The case was decided in Mary's favor in 1667.[60] As the bridge works tarried in recommencing operations, however, the New River invaded its supply zone and took up some of its former customers. Mary Morris petitioned the Crown to force the New River to stop encroaching on its territory, and even to take up the pipes it had laid.[61] As with the Durham Yard Waterworks, there is no record that the Privy Council was in any way open to this sort of argument. In any case, the New River supplied that part of London in the years to come, and competition between the two companies became a fact, although they mitigated it with agreements not to transfer customers with unpaid debts. In all these cases, the Crown favored the New River's right to expand. To what extent it was motivated by political influence as opposed to a principled view about the desirability of competition is not clear. In at least one instance, however, the Privy Council specifically referred to the New River's yearly payments to the Crown.

Improving and Joint-Stock Companies, 1660–1700

The formation of new water companies, a process that restarted in 1655, accelerated after 1669 before slowing markedly after 1705. As with the first period of company formation, this new era followed broader trends in the English economic and business environment. Most notable was a renewed interest in projects originating from 1650, in addition to booms in patenting and especially joint-stock company formation, beginning around 1680. From the end of the Civil Wars, and even before the Restoration, there was a return to the increasing enthusiasm for "improvement" in English society, one that echoed the late sixteenth century's age of projects. From 1650 onward, many proposals were put forward for improvements to the economic, social, and political life of the nation. The intensifying competition with the Europeans in overseas trade, especially the Dutch, was a spur to ideas about improving England's competitive position. This competition was particularly intense with the outbreak of the Anglo-Dutch wars during the Commonwealth in 1652, continuing sporadically until 1674. The Restoration of 1660 did not abate the enthusiasm for commercial improvement in an expansive list of activities. There was a particular emphasis placed on improvements that could help the country gain an edge in international trade and that combined with domestic advantages. The types of projects suggested and sometimes undertaken were wide-ranging, including, for example, beekeeping, land registries, enclosing land, and clearing wastes. Many of the proposals could be quite wild, such as one request for a patent for a powder that "being put into fair water, beer, ale or wine immediately turns it into very good writing ink."[62] Others, such as mine and land drainage schemes, or new mercantile companies such as the Hudson's Bay Company, were quite successful. Another important group was in building construction, the most prominent member of which was Nicholas Barbon in London. Daniel Defoe also joined in the spirit of the times, and his first notable publication was *An Essay upon Projects* of 1697, which included a number of proposals, such as a society to improve the English language.[63]

Despite the projecting enthusiasm, there were serious limits placed on what the projects could achieve. Many of the developers hoped for the state's support but were disappointed because of notable restraints on the state's ability to act. On the one hand, political disagreements held up many plans, such as the widespread use of acts to enclose land to increase agricultural productivity. On the other, the state's own powers were constrained by the limitations and unpredictability of the central government. It was restrained administratively because of its small size, with few people directly in its employ. More importantly, the old conflicts between Crown and Parliament reemerged.[64] Although there was relative calm between the rival power centers after the Restoration, tensions mounted again in the 1670s and 1680s. Charles II repeatedly prorogued Parliament before the end of a session to prevent its continued meeting, making acts more difficult for promoters of projects to win. They turned once again to applying for patents, producing an increase in awards in the 1680s.[65] Although Parliament was meeting again in the later 1680s, the patenting trend became a surge, with a total of sixty-one granted in the years 1691 to 1693, a number not matched again until the 1760s. This crest in patent awards was associated with the contemporary joint-stock boom. Most patents issued during the boom of the 1690s were for military-related industries, which were buoyed by a new war, this time initiated by William III's against France. Patents were still not yet being granted exclusively for new inventions, and they remained attractive for water companies because they could give them the right to build waterworks over the objections of locals. What was different from earlier patents was that patentees began using them to publicize their projects. Patentees liked to trumpet the royal approval the patents were claimed to signify.[66]

A further decisive impetus for the creation of new water companies was the financial revolution, which was a series of related developments in the English business and financial environment from 1685 onward.[67] The focus of these changes was particularly in the national debt as issued by Parliament following the Glorious Revolution of 1688 and its trade in the secondary market. Other elements included the new Bank of England and the growth of a secondary market in trading in this debt as well as in stocks. Historians have usually associated the financial revolution with the Glorious Revolution of 1688, when the Catholic James II was deposed and replaced by Parliament with the Protestant William III and his wife Mary, who was James's daughter. Most notably, North and Weingast famously argued that the Glorious Revolution was a turning point in the economic development of Britain. They claimed that the decisive transfer of political initiative from the Crown to Parliament created a propitious environment for business activity because political and legal control now lay with Parliament, which tried to ensure that the state respected property rights. Previously, such rights were subject to the whims and vagaries of the Crown and were therefore less secure. Government debt, one of the central elements of the financial revolution, was a key example. Before 1688, state debt was issued by the Crown and could be repudiated more easily on a royal whim, such as Charles II payment stoppage in 1672. After the revolution, debt was issued by Parliament, and its members ensured that it was paid because many parliamentarians were part of the same group that was owed those very same debts. The public, therefore, had much more confidence in the state and its financial commitments. A series of innovations followed, and these laid the foundations of a new market in government debt from 1693. The expansion and liquidity of the market were such that despite enormous borrowing throughout the wars of the eighteenth century, the interest paid on this debt decreased markedly, down to 3 percent by the mid-eighteenth century. The growth and liquidity of this market were aided by the companies that specialized in government debt, notably the Bank of England. They held huge sums of long-term and difficult-to-transfer government debt, while their own shares were easily traded.[68]

More recently, elements of this thesis have been challenged. For example, some historians have argued that secure property rights were present before the revolution.[69] Julian Hoppit has shown how they were at times undermined afterward, such with the expropriation of land for canals.[70] In addition, the interest rates charged to Parliament were not significantly lower than what unreformed sovereigns could get elsewhere in Europe for many years.[71] Anne Murphy has also pointed out that due to difficulties in collecting taxes, the parliamentary government delayed debt payments in 1696 and 1697, and again between 1708 and 1710, similar to Charles's stoppage a few decades earlier.[72] Finally, some elements of the changes included in the financial revolution had already been in place before 1688, such as a resurgence in joint-stock company formation. Nevertheless, despite historians lessening the centrality of the Glorious Revolution in the evolution of financial revolution, many of the financial revolution's achievements remain in place, especially when seen in a longer perspective. For the purposes of the water industry, these included the post-1685 joint-stock boom and the flourishing of London's stock market.

The late seventeenth century witnessed a dramatic increase in the number of joint-stock companies. There were about fifteen of them in England in 1685, and a further one hundred were established to 1695.[73] The reasons for this boom included a generally improving investment environment after the Great Fire with low interest rates, along with growing wealth in the city.[74] The international situation also helped. William's war with France created opportunities for English merchants because French ones were denied access to English ports, shifting some of their business to the native trading firms. The end of the French trade also fostered domestic industries, especially in luxury goods traditionally imported from France, such as linens and silks. Finally, the war also stimulated demand for materiel, and, as indicated earlier with patents, many of the new joint-stock companies were created to meet the conflict's needs. The boom in company formation ran to approximately 1693 and diminished thereafter. The slowdown in the market was caused by a variety of reasons, including the notoriety that stockjob-bing and projecting had gained by some projects going bad, shipping losses during the Nine Years' War, a slump in credit with the prolongation of the same war, and the poor quality of coinage. In addition, the rise of trading in the public debt in 1694 and the associated moneyed companies crowded other possible investment opportunities.[75] This last shift occurred especially after 1700 when the three moneyed companies—the East India Company, the Bank of England (founded 1694), and the South Sea Company (founded 1711)—eclipsed the rest in terms of capitalization. These last two companies especially, but even the East India Company, were focused especially on holding government debt, despite the nominal interest in trade for the South Sea Company. They owned 39 percent of outstanding debt in 1714.[76] The bit-ter end to the boom prompted the likes of Daniel Defoe and others to attack the mendacious manipulations of traders.[77]

The formation of all these new companies also saw the emergence of the London stock market. Although there was as yet no formal stock exchange, after 1685 securities were exchanged especially in the coffee houses around the Royal Exchange, a merchant's market established in 1571. The growth of the stock market included the appearance of financial professionals, nota-bly stockjobbers, who traded on their own account, and stockbrokers, who acted as agents for shareholders. Trade volumes were relatively low before the boom of 1690.[78] Although the stock market grew in London, water com-pany shares traded thinly during much of the eighteenth century with the New River held by very few people, and the rest too small for active trading.[79]

All these developments, together with the new enthusiasm for improve-ment and the financial revolution, had observable effects in infrastructure construction in England, such as in river improvements. After 1660, projec-tors proposed and Parliament was often willing to grant river improvement acts, which typically allowed a group of people to dredge, widen, canalize, or otherwise improve the navigability of existing rivers, also giving them the right to levy tolls.[80] This marked the beginning of a long growth of river trans-portation improvements in England, culminating in the canal construction beginning in the 1760s.[81] The broader trends also had effects on the London water industry from 1666 onward. New entrants to the industry appeared, both as competitors to the existing companies and in areas not well served by them. Entrepreneurs likely detected an opportunity in water supply from the New River's lucrative growth and subsequent supply difficulties. As with the first wave of water companies, some form of collaboration with governments, at either the national or the local level, was essential for enabling the project because of the issues of rights to water and access to land to build a network of pipes. Once again, the support these companies received took the form of patents and charters that granted these rights, as well as agreements with the City and other local entities for access to city streets.

One of the first of the new companies, the Shadwell Waterworks, was formed in 1669. It was situated to the east of the City where neither the New River nor the LBWW was supplying water. Its founder was Thomas Neale (1641–99), a serial projector who was emblematic of this new age of improving projects. The waterworks was the first of at least thirty-nine projects he promoted between 1669 and 1696, in fields as wide-ranging as manufacturing dice to raising wrecks off Ireland to running lotteries. Neale was born into a family of landowning gentry in Darnford in Hampshire. He was educated at Cambridge in the late 1650s and would later be admitted to Middle Temple as a lawyer in 1672. He married a rich widow in 1664 and settled at his family's extensive estates in Hampshire during which time he held a number of local offices, such as high sheriff in 1666. He was elected an MP in 1668, a position he held for twenty years, representing seats in Hampshire. While in the Commons, he served on many committees, including ones related to finances, a position that allowed him to win favor in the royal court. His wealth and influence were such that by the early 1670s, he was regarded a "person of vast estate and of great Interest as well at court as in the City and country."[82] During his years as MP, he repeatedly voted with the Crown during the exclusion crisis between 1679 and 1681 that raged over whether the Catholic Duke of York, the future James II, could succeed his brother to the throne. This closeness to the Crown paid off in his ability to get patents for projects, the first of which came in 1676. Over the years, he applied for ten English and Irish patents, of which he got at least five. Besides the waterworks patent, he was awarded ones for the production of tin, steel, and brass. He also received positions that signaled royal favor, such as being made the master of the mint in 1678.[83]

Neale planned to make the waterworks part of a larger project to develop Shadwell, a neighborhood to the east of the City. Much of the land there was owned by the dean of St. Paul's Cathedral, and Neale leased it in 1669. He succeeded in getting Parliament to create a new parish for the area and went about erecting buildings for new residents, including a church, brewhouse, market, yards, and waterworks in 1669. The waterworks drew water from the Thames with pumps powered by four horses. For the first years, the waterworks supplied only Shadwell, but in 1673 Neale considered extending his pipes into the City to the east. He applied for a patent to do so, and Christopher Wren, the surveyor general then responsible for much of the reconstruction of London following the fire, reported that it could usefully serve areas until then "not yet well supplied with water." He was given a patent to operate in the City in 1675 but still needed permission from the City's aldermen to access the streets, which he received.[84] After Neale augmented the waterworks' capacity, he applied for and received a further patent in 1680 from the king that allowed him to split the ownership of the waterworks into thirty-six shares, some of which he sold. He finally incorporated the Shadwell Waterworks through act of Parliament in 1692. The company lasted until 1807 when it purchased by the East London Waterworks.[85]

The York Buildings Company was another enterprise formed in the years immediately after the fire, when James II granted a patent to Ralph Bucknall and Ralph Wayne in 1675.[86] The patent allowed them to construct their waterworks on the site of the York House estate located on the Thames south of the Strand near Charing Cross, which was then being parceled off (fig. 3.4). Perhaps wary of the competition that had prompted the removal of some of the New River's pipes in its battle with the Durham Yard Waterworks, the patent enjoined the York Buildings Company and the New River to avoid such tactics.[87] The works used a horse pump until 1683 when it burned down. It recommenced operations in 1685 and, riding the joint-stock boom, got an act of incorporation from Parliament in 1691, at which point it had seven proprietors. The act also contained a clause safeguarding the New River rights to operate in its area. The company served the mostly wealthy neighborhood that surrounded it. It had a long subsequent history as a corporation, but one that largely pursued commercial activities other than water supply. It was later discovered to be involved in fraud and was eventually prosecuted for engaging in activity unrelated to its original charter.[88]

Another company with a similar history was the Millbank Waterworks established by Michael Arnold, a brewer in Westminster, together with four co-owners beginning in 1673. The company received two letters patent, one in 1675 and another in 1678, giving rights to supply much of the local area being the parish of St. Margaret in Westminster. It was also powered by a horse engine.[89] The company was purchased by the Chelsea Waterworks in 1727.

The Shadwell Waterworks, York Buildings Company, and Millbank Waterworks were all formed by 1685, before the joint-stock boom got underway. They all initially relied on patents but then participated in the joint-stock

Figure 3.4. The York Buildings Company water tower. *The City of Westminster from Near the York Water Gate* (1746), by Canaletto, Yale Center for British Art, New Haven

boom by later becoming joint-stock companies. The boom then fostered the birth of even more water companies, but it was not the only stimulus. The financial bankruptcy of the City in the 1690s also contributed. The City's bankruptcy prompted it to transfer more of its own water rights to private interests, as it had done with the earliest companies. In addition, it made a particular effort to avoid having these rights fall into the New River Company's hands. The City was driven to this recourse out of financial need, a reason different from what had motivated it several decades before. Originally, it had been London's inadequate water supply that prompted it into action. By the 1690s the situation had changed. Water supply was passable in many areas with water companies willing to meet demand, but the City was in an increasingly precarious financial situation. The roots of this lay in its handling of a fund for the support of the orphans of its citizens, paid into by the estates of deceased people leaving children. Due to heavy expenses stemming from wars and especially the Great Fire, the City had dipped into this fund. By the 1680s the City's liability to the orphans was so great that its interest payments to the orphans fund constituted 75 percent of its expenditures.[90]

By the 1690s, the City was effectively bankrupt and, as it did during James I's rule many years before, it sought extra revenues, including funds gained by selling water rights in 1692. These measures did not save the City, and its debts had to be restructured by Parliament in 1694, but it did create new water companies.[91] The act restructuring the debt specifically protected the rights of the New River, the LBWW, and the York Buildings Company. It furthermore forced the City to put all income from water rights toward paying creditors.[92] As a consequence, the City was impelled to farm out all other water rights to contractors.[93] In contrast to the earlier cases, therefore, the City did not give away the water rights freely and expected large sums in return, placing the new water companies under financial pressure that hobbled and even ruined many of them within a few years. Indeed, most of the people who signed leases for water rights with the City at this time became entangled with it and its orphan creditors in lawsuits as the expectation for returns were inflated during the heady joint-stock boom, leading to unrealistically expensive leases.

One of the new companies created within the ferment of joint-stock fever and the sale of water rights was the Hampstead Waterworks, incorporated in 1692 with six hundred shares nominally valued at twenty pounds each. As with the rights to the Hertfordshire springs the City gave to Myddelton, the City had been granted rights to water from Hampstead, a hill to the northwest of the City, by parliamentary act in the sixteenth century.[94] The City may have drawn some water from Hampstead before the late seventeenth century, and the rights were still under its control in 1692 when it signed an agreement transferring them to a few London merchants and bankers, including William Paterson, the founder of the Bank of England.[95] Unlike with the New River, which got the rights for free, however, the City expected and needed rent, charging £80 per year for thirty-one years, in addition to an initial fee of £200.[96] In a separate agreement, the City leased water on the south side of the Thames to Patterson and others for ninety-nine years for £550 plus £250 annually.[97] The City kept to these terms and pursued the proprietors, even in court, when they failed to pay.[98] The water enterprise to the south of the river was abandoned, but to the north the Hampstead Waterworks survived but did not thrive. It became embroiled in lawsuits with London residents, who perhaps resenting the loss of their own access to Hampstead water, made the companies access to some streets difficult.[99] The New River's supply problems in the West End gave the Hampstead Waterworks scope to poach its customers. As a 1701 letter from Evan Jones to Robert Harley, an MP and New River proprietor related by marriage to the Myddelton family, described, the inhabitants of the West End to the north of Piccadilly currently served by the New River were "so uneasie at the ill servitude that they are inviting the Hampstead Waterworks to lay in pipes in those streets, which if not speedily prevented will be some hundreds pounds per annum out of the Company's present income in those parts only."[100] The Hampstead Waterworks managed to capitalize on this opportunity only to a limited extent, and when some of the originals owners eventually faced financial difficulties, the lease was transferred to new owners in 1715. They also struggled and by 1737 owed the City over four years of rent.[101]

Another 1692 case was that of the company called the City Conduits (or the Marylebone and Paddington Conduits). Some London residents led by Thomas Houghton and John Tyzack signed a couple of leases with the City for forty-one years for all the waters the City had rights to north of the Thames that were as yet unleased.[102] The lessees contracted to pay £700 per year plus £2,650 initially, as well as to supply some conduits at no charge. The entire project soon descended into bitter recriminations between the City, the water farmers, and Robert Aldersea, a plumber who had had a contract to repair the City's pipes feeding one of the conduits. The water being supplied to the conduit was nowhere near what the contract had stipulated, and the farmers placed the blame on Aldersea's work.[103] The affair ended with years of litigation, eventually landing some of the proprietors in debtors' prison in 1704.[104] By 1703, however, the City had cancelled the lease and signed another contract with a London goldsmith, Bartholomew Soame, and other investors.[105]

The sale of the conduit water rights to Bartholomew Soame was tied to a simultaneous attempt by his son Richard and other investors to buy the London Bridge Waterworks and merge it with the conduit waters.[106] Richard had been in negotiations with the owners of the LBWW for some time and reached an agreement to purchase the works in 1701 for thirty-six thousand pounds.[107] For the sake of comparison, the Royal Navy's largest ships cost around thirty thousand pounds to build at about this time.[108] The competitive pressure from the New River apparently prompted the Morris clan to sell.[109] The LBWW was then constituted as a joint-stock company with five hundred shares, which were later divided into fifteen hundred.[110] Thomas Morris, acting for the company, negotiated with the City to lease the fourth arch of the bridge, closer to the center of the river where the flow was greater. The City consented to this lease on the understanding that the LBWW would also lease the conduit waters as well.[111] This condition was important for the City because the lease of the arch was for a nominal amount but that of the conduit waters was for an initial payment of three hundred pounds, and seven hundred pounds per year thereafter. The difficulty was, however, that the parties in negotiations with the City for the two halves of this deal were not the same. One was Thomas Morris, who acted for the new owners although he had sold out by then. For the conduit waters, it was Bartholomew Soame with a few other investors. Once the City had approved the lease of the bridge arch but before the lord mayor had signed it, Soame and company became very reluctant to sign the lease for the conduit waters. The result was a complex web of lawsuits. On the one hand, the conduit water contractors had already given a bond for ten thousand pounds binding themselves to the lease, and the aldermen decided to execute the lease as originally drawn up. The matter ended before the courts, with the lessees liable for sixty-nine hundred pounds.[112] On the other hand, the City explored its legal options in withdrawing from a lease of the arch it had approved but which had not been finally signed by the lord mayor. After various prevarications and delays, the City finally did lease the arch to the LBWW.[113] As part of the deal, Morris had agreed to lower the rent charged to its customers to twenty shillings per house.[114] The City was also worried about collusion between the LBWW and the New River.[115]

The New River did indeed become involved in the lease of the arch but not as the LBWW's ally. As had been the case earlier in its various petitions to the Privy Council regarding other water companies over the preceding years, the New River was willing to try to use political connections against its opponents. After the City had approved the new LBWW lease in 1701, the New River tried to prevent the deal from going ahead, arguing in 1702–3 to the Privy Council that, because blocking up the fourth arch would create a nuisance for boats on the river, it should therefore be stopped.[116] The City was not happy with this attempted interference on the New River's part and objected to the Privy Council, adding the observation that the New River charged unreasonable fees to the citizens of the City.[117] The New River had miscalculated, however. Unlike in the earlier cases, it did not prevail with the Privy Council, and the lease went ahead. The City was a more influential opponent than the small water companies it had been in dispute with earlier.

The net result of all these events was that the LBWW had passed out of the hands of the Morris family and had become an unincorporated joint-stock company. The works themselves were extended by the addition of a new waterwheel under the fourth arch of the bridge.[118] The conduit waters were soon cast adrift by the London Bridge Waterworks, left as a liability of the partners who had signed the lease with the City, and who suffered enormous personal losses in the judgment against them. The conduits company survived the lawsuit as a separate entity and was probably sold on. Records of it collecting rents from customers exist to 1771, by which point it was back in the City's possession. It was, however, tiny. It had 188 customers in 1747, 162 in 1757, and 135 in 1770.[119]

There were some new companies in the joint-stock boom that did not get their water rights from City, as had been the case with the York Buildings Company, the Millbank Waterworks, and the Shadwell Waterworks. These new companies had fewer financial and legal difficulties but were generally small. Hugh Marchant, an entrepreneur who had tried unsuccessfully to get a charter for a water company from Parliament in 1691, managed to get a patent in 1694 that allowed him and his partners access to sewers to drive waterwheels to pump from the Thames.[120] The commissioners of sewers for the area allowed them access to the sewer under Harthorne Lane off the Thames where they had leased a lot.[121] The following year, they petitioned the king once again for added powers. They had realized that they could not lay pipes through the city's streets without some form of governmental authorization.[122] Marchant was opposed, however, by the City, which by this point thought there were too many pipes under the streets in the area he was proposing to build, and this would harm its own water supplies.[123] Marchant prevailed, and his water pumps were finally built. Marchant's company (sometime called the Marylebone Waterworks) survived for some time, although it was unremarkable. When, in 1776, the commissioners of sewers realized that the company was no longer supplying any water to anyone, but only grinding corn, the patent was revoked.[124]

The result of this joint-stock boom was significant for the New River because of the competitive pressure it fostered. Even before many new companies emerged to trouble it, the New River's rapid growth had been too fast for the company's own good, and it proved unable to keep up with the demand for water from all the new customers it was taking on. In 1682 John Aubrey recorded that "London is growne so populous and big that the New River of Middleton can serve the pipes to private houses but twice a weeke."[125] Previously, it had supplied water three times a week to most of its customers. The situation continued to deteriorate, and in 1688 the New River directors ordered that a survey be made of all its customers, and report if "any water [can] be . . . spared."[126] The City also became involved, and the aldermen resolved to investigate not only "the failure of the new river water in diverse parts of this city" but also how to improve the situation.[127] From the City's point of view at least, if not the New River's, the solution was to bring in more water companies. People in the areas to the west of the City were also disaffected, and they too looked to other companies, complaining of poor supply from the New River. One letter described how the New River's supply to that area was weak, with its mains freezing before any other company's in cold weather.[128] It was to serve the wealthy West End that most of the new companies such as the York Buildings Company, the Millbank Waterworks, the Chelsea Waterworks, and others were created. To make matters worse, the New River's problems were really only beginning in 1688. Its supply difficulties were not immediately solved and would not be until after 1710 through a technical overhaul of its infrastructure. In the meantime, the New River lost customers to the LBWW and the newly emerging competitors.[129] Its revenue decreased between 1691 and 1710 by £9,679, and in some areas it "met with the strongest competition," as described by one of its shareholders.[130] Its dividends collapsed from their 1692 high of £255 to £113 in 1696, recovering somewhat in the following years to more than £200, before falling into a trough of around £170 between 1702 and 1713.[131]

New Attempts, 1700–1730

By 1705, most of the companies that would serve London north of the Thames during the eighteenth century were in existence, with two exceptions: the Chelsea Waterworks, incorporated in 1722, and the West Ham Waterworks, created in 1743 and incorporated five years later.[132] Formation of the Chelsea Waterworks took place during a revival in the joint-stock company following its fall from grace with the end of the joint-stock boom before 1700. From that time to 1717, only four new companies were formed.[133] Activity was rekindled after that, and a few new insurance companies were incorporated, and some old disused charters were repurposed. At the end of 1719, a new boom began. Shares in the moneyed companies soared, with the South Sea Company rocketing 820 percent in a few months. Smaller companies rode their coattails, and speculators saw opportunities for forming new ones. Within a year, hundreds of new companies were formed, and as had been the case a couple of decades earlier, some were based on wild and delusional ideas. Most of these bubble companies were unincorporated.

Parliament intervened in the bubble in 1720. In an effort to push money into the South Sea Company's stock, which was refinancing the national debt, it passed what came to be known as the Bubble Act. It forbade the formation of joint-stock companies except by a parliamentary act of incorporation. The hope was that all the investment pouring into unincorporated companies would now move into the South Sea Company, among other corporations, helping the debt conversion. In the event, the bubble soon burst, causing losses, scandals, and recriminations. Moreover, despite the letter of the law, the Bubble Act slowed but did not inhibit the formation of unincorporated companies over the eighteenth century since its original purpose had been for inflating the South Sea Company's stock, an event whose relevance soon faded.[134] For the purposes of the water industry, this episode had two effects: it spurred more interest in creating companies, and henceforth new companies sought acts of incorporations straightaway.

It was not only financial conditions that prompted new water supply schemes. The continuous growth of London, especially in the West End, inexorably increased demand for water. In 1717 the Hampstead Waterworks increased the size of its reservoirs, following a particularly dry year when its supplies had been problematic.[135] This led Thomas Archerley, a surveyor of "lands, mines, levels, and throwing out waters," to explore the possibility of getting water from the northwest of London, from a group of springs and rivers there, chief among them the River Colne. This watercourse had figured in Ford's and Roberts's proposal in the seventeenth century. Archerley surveyed land and laid out a possible course, finally joining with others to attempt to set a new water company afoot. They tried to approach Parliament in 1718 to get an incorporating act but failed. They tried again in 1720. A petition to introduce a bill was made in February, with the petitioners speaking of supplying the "very greatly increased" inhabitants of London. The question was sent to committee, when favorable testimony from various luminaries, including Edmund Halley, stated that the project was practicable.[136] The bill was then introduced to the House, and the proposal faced repeated vehement opposition in subsequent sessions from mill owners, concerned over water losses for milling, and landowners, such as the earl of Essex, across whose land the aqueduct would run. Support came from some inhabitants of Westminster, who stated that the water proposal "will be not only useful and beneficial to the whole town, but is necessary, and much wanted in those parts." The opposition, however, was too strong, and the bill stalled in the Commons and died.[137] The supporters of the bill accused the New River Company of stirring up the opposition among mill owners.[138] They tried to bring in a new bill in early 1721, but the petition was simply rejected.[139] Archerley and his companions, however, were determined and tried again in Parliament in 1722–23, and once more in early 1724, before making one last attempt at the end of 1724 by approaching the Privy Council first for support, all to no avail. The usual vested interests of mill owners and landowners prevailed, probably with some help from the New River, which had never been shy about lobbying against potential competitors.[140]

A second bill was brought in in January 1721 by a different group hoping to incorporate a company to bring water "from one or more streams or rivers that run by or near the village of Drayton," especially the Cowley and Heatham Streams.[141] Like the Colne proposal, it was sent to committee, before which experts were called to testify on its merits, including the natural philosopher John Desaguliers, who was later involved with the York Buildings Company. He testified that the proposed channel could draw three times as much water as the New River to a reservoir near Hanover Square. The water could be pumped from there into a higher reservoir and supply 90 percent of the houses in the area. Other "mathematicians" also testified that the lay of the land was adequate to get the water to London.[142] After various delays, the bill attracted strenuous opposition from many vested interests, including barge owners, mill owners, and many towns and landowners along the rivers to be affected by the proposal. Even the City of London and the City of Westminster both petitioned against the bill, arguing that it would harm navigation on the Thames.[143] The River Lea trustees added their voices to the opposition, not because it affected them directly, but because the New River already diverted too much water from the Lea and would be emboldened to take more should the new proposal come to fruition.[144] Even Oxford University joined the petition against. Besides a few barge owners who thought the Thames would be deepened, the only group to lobby in favor of the bill was the Hand-in-Hand fire insurance company, which claimed that because of inadequate water supply, not only "great losses have happened, but the damages sustained by their office have greatly increased."[145] The consideration of the bill dragged into 1722 when it was finally defeated in June. The ferocity of the opposition was such that Frederick Clifford observed in his history of private bill legislation that "up to this time there is no record in Parliament of a private Bill which had aroused such determined and widespread antagonism."[146]

The defeat of these two bills in short succession left the New River as the only company drawing water by aqueduct from north of London. All the rest functioned by pumping from the Thames. Given the very broad and fierce opposition the two canals proposals had aroused, it was doubtful that any further such projects could have succeeded in getting an incorporating act in the early eighteenth century. Even the New River itself had not received its powers directly from Parliament. Rather, they had been given to the City by Parliament and then passed on. The New River had failed on multiple occasions to get an act from Parliament authorizing its status as a company running an aqueduct from the north. The only companies that had succeeded in getting acts did so by limiting their ambitions to pumping from the Thames.

The opportunity to supply to the West End remained, however, and the joint-stock boom still had enough strength to throw up one more proposal. This was the Chelsea Waterworks. Learning from the defeat of its predecessors, its promoters suggested the Thames as its source of water. The company was founded in 1721 by group of investors led by John Fane (1685–1762), a lawyer, army colonel, and MP for Kent, as well as Richard Molesworth (1680–1758), MP for the Irish Parliament and a lieutenant colonel in the army. The promoters presented their petition to Parliament in January 1722, being careful to minimize potential hostilities from vested interests by claiming that the project would not harm the Thames navigation. The bill was sent to committee, and Lord Molesworth (Richard's father as it happened) reported favorably on its behalf when the bill returned. The promoters had obtained the consent of the dominant local landowner, Lord Grosvenor, beforehand, so no objection was raised from that quarter. The bill passed in a month's time, with not a single counterpetition presented.[147] The Chelsea Waterworks was incorporated with a nominal capital of forty thousand pounds by charter granted in 1723, a year after the act passed. It issued two thousand shares with a nominal value of twenty pounds each. The original selling price was likely twelve pounds each, with two-pound calls raising the invested capital to the legal limit by 1726.[148] Unlike the New River, the Chelsea had a relatively wide base of shareholders. There were 141 in 1730, most of whom held five shares.[149]

The new corporation soon acquired a site on the Thames in Pimlico. There, the company built channels connected to the river. These were allowed to fill at high tide, and when the tide was low once again, some of the water was let out and used to turn waterwheels to pump water up to reservoirs in St. James's Park and Hyde Park, sites it had acquired by 1726.[150] From there, it supplied water to Westminster and the West End, as well as to St. James's Palace and Kensington Palace. At first, the Chelsea grew organically, getting customers on its own, and had about nine hundred in 1727.[151] It was able to grow in part because a spring in Hyde Park failed in 1725, and locals were seeking new sources.[152] The Chelsea Waterworks then made major purchases, beginning in 1727 with the larger Millbank Waterworks, which at that point had thirteen hundred customers.[153] The Millbank Waterworks pumped water from the Thames with a horse engine. This acquisition and continued internal growth took the tenant roll to about thirty-two hundred in 1729.[154] In 1732 the company made another acquisition when it took over a reservoir located north of Cavendish Square from the York Buildings Company, and acquired a number of tenants in the area as well. York Buildings sold the reservoir because it found the coal needed to power its Newcomen steam engine, installed in 1726, was simply too expensive and gave it up; it was no longer able to supply the neighborhood.[155] All these purchases had exhausted the company's initial capital, and it could raise no more through share sales because of its charter-imposed limit. The company issued bonds as a short-term solution but petitioned Parliament to extend its capital by thirty thousand pounds. This was granted in 1733.[156] The company sold two thousand new shares for ten pounds each by 1735, paying off all debts.[157] The remaining shares were never sold.[158] The fresh capital allowed the Chelsea Waterworks to make another major purchase in 1738 when it bought the Hyde Park Waterworks, with approximately four hundred customers, taking its list to forty-six hundred.[159] It had overexpanded, however. During the winter of 1739–40, there were supply failures and bitter complaining from its tenants, prompting customer defections. The directors added two Newcomen steam engines to increase supply. This stabilized the situation sufficiently that the complaining largely died down, and the company stopped losing customers. It did not, however, thrive. Its eight shillings per share dividend, which it started paying in 1737, was halted in 1742, and not reinstituted until 1754, at which point it was six shillings. It rose only in 1771 when it went back to eight shillings. It was 1797 before it moved again, this time to ten shillings (see fig. 5.1). The number of tenants it held was also quite stable, although no exact figures are available from 1738 until the early 1800s, at which point it had eighty-three hundred. This indicates very slow growth, almost stagnation, during most of the eighteenth century.[160] The Chelsea Waterworks had acquired more customers from 1722 to 1738 than it did in the following seventy years. During the long period of no dividends, the company was financed through debt, accumulating more than four thousand pounds by 1747, which was slowly repaid by 1753.[161]

More is known about the Chelsea Waterworks than the other companies founded at this time, in part because it was larger and kept better records. Compared to the New River, whose constitution was sui generis, the Chelsea's was more in line with customs of other joint-stock corporations. It was governed by thirteen directors, a governor, and a deputy governor, all elected at an annual general meeting of the shareholders. The directors appointed the other officers and positions within the company. Extraordinary meetings of shareholders could be called by the governor or by five shareholders holding at least ten shares. Voting rights were capped at three: a shareholder with five shares of twenty pounds each had one vote, ten shares conferred two votes, and twenty or more shares gave three votes. Shareholders with shares in other London water companies were barred from voting.

Conclusion

By 1725, the London water industry (on the north side of the Thames) had acquired business outlines it would hold until 1800, with the exception of the West Ham Waterworks founded in 1743. The major companies were in place, and their zones of supply roughly settled. Growth would certainly continue as the metropolis grew and more houses were built, but it was the New River that picked up most of them (see chapter 6). In terms of business form, all the large companies and most of the small ones were now joint-stock companies, and most were corporations with the notable exception of the LBWW, which remained unincorporated, with seemingly no disadvantage. The water industry had followed the broad contours of the evolution of joint-stock business over this time. Before 1650, all joint-stock companies were corporations, of which there were very few. For the water industry, the sole example was the New River Company. By 1720, after the late seventeenth-century joint-stock boom and the emergence of unincorporated joint-stock companies, the New River was no longer exceptional from the point of view of legal form. It stood apart only in the unusual dual-class share structure it had inherited from its alliance with James I. The model of water supply managed by joint-stock companies, particularly corporations, was firmly established.

Another feature of this period was the consolidation of the model of water supply based on for-profit businesses running network infrastructure. Around 1625, some doubts may have existed given the difficulties in holding on to customers and the reliance of the first water companies on government support in the form of loans and investment. The marquee company, the New River, was not yet paying dividends. By 1720, and even by 1670, all doubts had been dispelled. The New River was booming. It had one of the highest stock capitalizations of any joint-stock company. The other companies, such as the LBWW, which was paying healthy dividends to the Morris family, were doing well, too. Many entrepreneurs were attracted to the model and piled into the industry, especially after 1670. They stoked fierce competition for some time, prompting bankruptcies, consolidations, attempts to use influence with the Crown to control competitors, and even underhanded tactics such as removing other companies' pipes. By 1700, the basic model of water supply was firmly set, and it entailed private companies operating a network infrastructure feeding houses directly. The City had implicitly acquiesced to the advent of this model by allowing its old conduit infrastructure to degrade, especially after the Great Fire.

Finally, governments continued to loom large in the industry through the turn of the eighteenth century because royal patents and parliamentary acts continued to be possible means of getting sufficient rights to run a company over the objection of vested interests. The constantly fluctuating power dynamics, with wars, revolutions, displaced monarchs, and prorogued parliaments, made for a very uncertain field for water entrepreneurs to negotiate when choosing the best route. Most simply tried the path that was available at the time, seeking rights through patents during Charles I's personal rule or when Charles II repeatedly prorogued Parliament in the 1680s. Personal connections helped, such as Thomas Neale's solid support for the Crown aiding his bid to get his Shadwell patent. When the political winds shifted again, these entrepreneurs sometimes sought to consolidate their position, such as the many who got acts of incorporation from Parliament after winning patents in the 1680s. Some, however, were caught out by too rapid changes, most especially the Somerset House Waterworks, whose patent was granted by Oliver Cromwell, a source beyond the pale for the restored Charles II. Patronage was also important in the disputes between companies, which repeatedly appealed to the Privy Council to mediate. It rejected all claims to exclusive supply zones, though this was likely based less in a desire to preserve a competitive market than in the New River's influence over the body. The New River was a party to all the disputes before the Privy Council, and it served its interests not to be excluded from the neighborhoods in question, where its opponent was the one claiming exclusivity.

Patronage from the Crown or in Parliament, however, had its limits. Up to the 1740s, no company was able to get a parliamentary act allowing it to build an aqueduct from the north of London, despite the presence of abundant water there. Royal patents for this sort of aqueduct had been granted before the Civil Wars, but the curtailing of royal power had prevented such proposals from going ahead by patent alone afterward. All new proposals for such a scheme thereafter sought parliamentary approval. The vested interests resisting such a proposal, ranging from mill owners, to towns on affected water bodies, to landowners, were simply too powerful for the promoters of a new company to overcome. If they limited their ambitions to pumping from the Thames, resistance was much less, and the enabling acts and patents were within reach. This in effect protected the New River from a competitor using its own model. The company had been fortunate in acquiring the rights to build an aqueduct originally granted to the City, and in getting them while the Crown still had such powers to grant. The only exception to this state of affairs came in 1748 when the West Ham Waterworks was incorporated by parliamentary act and got permission to draw water from the River Lea. Its intakes, however, were on the river within greater London, relatively close to where the Lea meets the Thames.[162]

Finally, the City played a different role from what it had in the early years. Whereas the first companies received loans with easy and even generous terms, this was decidedly no longer the case after 1650. The City was willing to delegate more water rights, but in contrast to the earlier period, it demanded much higher rents. They were so high, that most of those who accepted them ended badly, bankrupt and sometimes in prison.

CHAPTER FOUR

A New Scale of Network
with the New River

The New-River . . . water . . . is dispersed in pipes laid alond in the ground
for that purpose, into abundance of streets, lanes, courts and alleys of this
City and suburbs of London; the great contrivance whereof all the citizens
have daily experience.

James Brome, *An Historical Account of*
Mr. Roger's Three Years Travels, 1694

By the 1690s, there was a tension within the New River Company. On the
one hand, its business model had turned it into a great success. It was generating large profits and had grown tremendously, inspiring imitators. On
the other, it had effectively outgrown its mode of operating, and its infrastructure was no longer adequate. It was having trouble serving all of its
customers. Moreover, its competitors were driving down the price of water,
causing the revenue it derived from the fines charged for new connections
to shrivel to close to nothing. By around 1700, shareholders were so disaffected with this state of affairs that they agitated for changes to restore the
company to the profitability and growth it had seen during the heady days
from 1660 to 1685 when it was unrivaled. Under pressure from its shareholders and competitors alike, the New River directors between approximately 1690 and 1710 commissioned a series of technical reports, seeking
recommendations for improvements to its infrastructure. Christopher Wren
and John Lowthorp, both members of the Royal Society and the authors of
these reports, diagnosed the company's problems and made suggestions that
were adopted slowly over the coming years in various ways. They both believed that a fundamental source of the New River's travails was that its infrastructure had grown haphazardly in response to demand and needed to
be rethought by taking into account the entire disposition of its works. From
this starting point, they recommended changes that would eventually set the
New River's distribution network on firmer footing, including the means to
improve its supply capacity and ways of delivering its water more reliably to

distant and more-elevated areas. Over the long run, the changes introduced from this time, combined with practices from the earlier era, slowly enabled the company to continue the growth that had hobbled it around 1690, while regaining its notable profitability. The dividends halted their decline in 1712 and then remained roughly constant until 1745. They then resumed growing and took off around 1775, increasing almost constantly until the nineteenth century. Although there are far fewer data available about share prices due to thin trading and the private nature of sales, share prices reflected this trend as well, particularly their appreciation in value after 1750, with a sharp rise after 1780.

The technical transition the New River executed around 1700 was slow but, in the long run, clearly effective. The company managed to stabilize its network when its wobbles had threatened its reliability and crippled its future growth. This stabilization of the New River's infrastructure in an enduring way was a significant technical achievement, one that was to be repeated with other large-scale infrastructure networks in the future. The ability to build stable large-scale and integrated networks would become a fundamental aspect of their design and construction. This is particularly true of tightly coupled networks, where the design and functioning of one element depend to an important degree on other elements within the same network. Historians of technology exploring modern technological systems, particularly of infrastructure networks, have made the construction and stabilization of these tightly coupled and integrated systems a central theme. Thomas Hughes's history of the development of electrical systems as networks has been seminal in this regard. He argued that successful builders of large technological systems not only designed, constructed, and refined networks by integrating their specifically technological elements in an overall design but also took account of nontechnological elements, such as politics and markets, as they deployed these networks on large scales. Hughes emphasized that the process of growth and scaling of these networks from local pilot systems to regional networks with tens of thousands of users involved more than simply extending the infrastructure. It also included finding solutions to the many new problems associated specifically with the scaling process, such as long-distance transmission with electrical networks. Managing load factors by balancing supply and demand was crucial to this stabilization. This early period of network construction was marked by trials with different standards and solutions before the system achieved a degree of maturity, consolidation, and even rigidity. For Hughes and others, this multifaceted and integrated design of complex technological systems emerged most clearly with the large-scale infrastructure networks of the nineteenth century.[1] The degree to which earlier networks were tightly coupled and integrated is not clear. For example, the use of specifications and standardized parts, both important means of stabilizing networks, was minimally present before the nineteenth century other than such basic standards as the width of canals.

This chapter examines the New River's technical transformation beginning in the late seventeenth century and extending over the eighteenth century from the point of view of the network design. It explores whether the construction and expansion of the New River offer any hint of an integrated design, or if it was an accidental network, built up by the accretion of pragmatic accommodations to immediate problems. Both approaches were present to varying degrees. The continuity in the basic design of the New River and the other water companies' infrastructure with the preexisting conduits and aqueducts is evident. Although new technology, notably waterwheel pumps and pipe-boring machines, catalyzed a new kind of water industry, there were nevertheless abundant commonalities between the conduit and water company models in the earliest years. The slow stepwise growth of the New River in particular, and then its rapid expansion after 1660, reduced these commonalities and effectively moved the company into a realm where incremental accommodations were failing to allow the continued growth of the company. As supply problems demonstrated, no longer would the ad hoc method of network building suffice. The directors realized that a new approach was needed, and from the 1690s there were signs of a more integrated approach to network construction, explicitly recommended by Wren and Lowthorp. This chapter shows how all this came about, and how the New River functioned over the eighteenth century.

The Scale of the New River

If the New River was a large-scale network, how big was it from the late seventeenth century onward? Was the New River Company truly different from other water networks? How did it compare with other companies, or the situation of water supply in other cities? A fuller discussion of this subject is given in chapter 6, but some observations are important here. From the one or two thousand tenants the New River had in its first decades, its expansion from 1660 rapidly pushed this figure up. From ten thousand tenants in 1670, it had seventeen thousand in 1683.[2] The New River's growth persisted after the turn of the eighteenth century as London's population continued to increase. Definitive numbers after 1683 are not available from the company's records until 1769, by which point the company counted around twenty-six thousand paying customers, or around 32,500 buildings based on an average rent of twenty-four shillings per building.[3] Although most of these were houses, other buildings, such as breweries, taverns, and stables, also received water. Between the two dates of 1683 and 1769 various figures were reported for the number of customers that show that a reasonable estimate for the number of buildings supplied was twenty thousand in 1720, and rising thereafter (see figs. 6.1 and 6.2).[4]

That the prevalence of water availability in London was remarkable by the early eighteenth century as compared to other European cities was reflected by comments from various observers. John Strype, who edited and rewrote John Stow's *Survey of London*, commented in 1720:

> There is not a street in London, but one or other of these waters runs through it in pipes, conveyed under ground: and from those pipes, there is scarce a house, whose rent is 15 or 20 £ per ann. but hath the convenience of water brought into it, by small leaden pipes laid into the great ones. And for the smaller tenements, such as are in courts and alleys, there is generally a cock or pump common to the inhabitants; so that I may boldly say, that there is never a city in the world that is so well served with water.[5]

Thomas Salmon made a similar observation in 1743:

> And as to water, no city was ever better furnish'd with it, for every man has a pipe or fountain of good fresh water brought into his house, for little more than the charge of twenty shillings a year, unless brew-houses, and some other great houses and places that require more water than an ordinary family consumes, and these pay in proportion to the quantity they spend; many houses have several pipes laid in, and may have one in every room, if they think fit, which is a much greater convenience than two or three fountains in a street, for which some towns abroad are so much admir'd.[6]

It was not only self-congratulatory Englishmen who made such comments; visitors also made remarks about water in the city. Voltaire, who spent three years in exile London in the 1720s, wrote many years later in 1767 to his friend Antoine de Parcieux that "I wish that all the houses of Paris had water like those in London do." Parcieux had been working for several years on proposals to improve Paris's water supply but with little success.[7] The Danish zoologist Johann Christian Fabricius visited London in 1782 and described how water "flowed through an unending mass of pipes through all streets, to almost every in this part of the city. There likely isn't another place in all of Europe so abundantly provided with water as London . . . Two water pipes run under the middle of every street underground, from which smaller service pipes connect to every house."[8] The German doctor Johann Grimm visited London in the 1770, and observed that "the lack of pure and good water occurs nowhere more frequently and easily than in large cities. London could have been an example of this, had they not found a way to avoid it. Besides various public fountains, there is no house where one cannot at will get water in an instant as soon as one has turned a faucet installed in the kitchen."[9] When first proposing a water company for Paris in 1777, Jacques-Constantin Périer, working with his brother Auguste-Charles, referred to London's water supply because of its "very great success."[10] In 1777 l'abbé Gabriel-François Coyer visited London and wrote that "the distribution of water is as extensive as the lighting. . . . Water . . . is shared through an infinity of pipes that take it into all the houses. There isn't one that doesn't have ready this element of basic necessity. The many men who used to carry it now do other work."[11] A former Prussian officer, Johann Wilhelm von Archenholz, wrote in 1785 that "every house is supplied with water by means of large pipes which run through all the streets."[12] Despite such positive assessments, opinions about the situation were not uniformly upbeat. There was much discussion within London about the quality of the New River's water. This is explored in chapter 7.

The pervasiveness of London's water supply from the early eighteenth century on was in contrast to supply networks to most other European cities and underscores its uniqueness as an early integrated network. Paris, despite repeated attempts at establishing water companies and expanding supply networks in this period, never moved much beyond the model of aqueducts or pumping mechanisms feeding fountains, with the widespread use of water carriers drawing water from them or from the Seine.[13] Even proposals for new companies envisaged water carriers going to fountains.[14] Similarly, no city in the Netherlands had a water distribution network before 1850.[15] As described in chapter 1, some German cities had pumping mechanisms feeding networks of pipes from the late Middle Ages, but large-scale expansion resembling what occurred in London came first in Hamburg beginning in the nineteenth century.[16] This was in large part because these cities were simply not as big as London and Paris. The situation in other British cities was similar. None were anywhere near the size of London, which completely dominated the country as an urban center. Edinburgh, for example, relied on public fountains and water caddies until at least after 1800.[17] For London's New River, however, building this network was not an unproblematic expansion of its earliest form. It required an elaboration of new means to stabilize the water network, especially from the late seventeenth century.

Wren's and Lowthorp's Reports

Although coming off a period of tremendous success, the New River Company was struggling in the late seventeenth century in new ways. Its troubles were far short of a catastrophic failure, but nevertheless they represented some serious problems. At the heart of these lay the company's inability to provide enough water to its customers, a situation that the company was well aware of. In 1701, for example, a report noted that customers around Swallow Street in the West End were so poorly served that they were entreating the Hampstead Waterworks to supply the area.[18] Another report commissioned by the directors in 1704 observed that some of the company's own employees were of the opinion "that in several parts of the town if the tenants were better serv'd their number would be encreased."[19] A further report from 1705 detailed that many complaints about "ill service" were made "both by [the company's] tenants and servants."[20] The supply problems were particularly acute in the areas west of the City where urban growth was the greatest. Between 1695 and 1705, six hundred new tenants had been added in the West End, and "great complaints were made by many of the tenants in [this] walke for not being served as they ought." In all the other areas put together, a total of five hundred new customers had been taken on.[21] The supply shortages in the West End were not, however, caused only because the company had taken on too many customers and overloaded its mains; the lay of the land exacerbated the situation. The elevation of the first receiving ponds at the New River Head were high enough to supply most of the City of London to the south and its immediate suburbs, but the areas to the west and south in Westminster, and especially in the West End civil parish of Marylebone, were more elevated, meaning that larger pipes alone would not solve the supply problem. Furthermore, as mentioned in chapter 3, the company's overexpansion was creating financial strain, too. The construction of new mains increased expenses, and despite acquiring new tenants, revenues shrank after 1700. All this, when combined with the new competition, resulted in weak profits and unhappy shareholders, who then put pressure on the directors of the company to continue increasing supply while restoring profit growth.[22] As a disaffected shareholder complained to the directors in 1710, there had been "great and unnecessary expences" over the previous twenty years as the company reformed and expanded its works.[23] Furthermore, the company even lost tenants between 1707 and 1710, causing its revenues to dip.[24] The directors, casting about for a solution during these years, felt they lacked the vision to do so and sought outside advice from Sir Christopher Wren and John Lowthorp. These two presented a number of reports around 1700 in which they provided some vision for an integrated design of the New River network.

Sir Christopher Wren (1632–1723) by the time he was consulted by the New River had had a long career. He studied at Oxford, where he became acquainted with some people who would later aid in forming the Royal Society in 1660. Interested in many subjects in natural philosophy, such as optics and astronomy, he also pursued technical experimentation in diverse areas including music, surveying, and printmaking. He worked as a professor of astronomy at Oxford and, after helping to found the Royal Society, came to the 'notice of the king, who began to consult with Wren. Wren had taken up an interest in architecture, doing some design work in Oxford, when he was appointed royal surveyor for the reconstruction of London from 1666 following the Great Fire, and then surveyor of the king's works in 1669, a post he held for forty-nine years. This gave him great scope for design work, including many churches destroyed by the conflagration, most notably St. Paul's, which was his crowning achievement. He also designed nonreligious buildings, such as the Naval Hospital and the Royal Observatory in Greenwich, Chelsea Hospital, the king's palace at Winchester, and Hampton Court Palace. He was an eclectic polymath and one of the leading figures of his age in England.[25] That the New River directors turned to Wren around 1700 indicated that they were seeking advice from someone with great experience in construction and design.

The New River's request to Wren for help was specifically to ask his opinion on how to improve supply to Soho Square in the West End. In his report to the New River Company, Wren echoed his vision for the reconstruction of the metropolis in the aftermath of the Great Fire in its emphasis on a reasoned overarching design, meant to do away with the chaos of the organic growth of the preconflagration city. His plan for the whole city had been infused with his views on experimental philosophy and was in the spirit of the Royal Society. He meant to re-create the city as the rational mind of the entire nation. According to Wren, urban space itself could be a support for experimental inquiry. In formulating this vision, Wren was taking a cue from Thomas Sprat's 1667 history of the Royal Society, in which Sprat claimed that London itself was a source of the precocious experimental philosophy emanating in the Royal Society. According to Sprat, London was a propitious environment for experimental philosophy because of its built environment, especially the coffee houses, created a social space that facilitated discourse by mixing people from all levels of society. It was in these spaces that the great questions of the day were debated, including philosophical ones. The rebuilding after the Great Fire offered Wren the possibility of bringing Sprat's vision of reasoned space to a new level in London. His plan featured broad spacious avenues and clear vistas provided by long straight streets. Just as the Royal Society was leading the new world of experimental philosophy, so too would this plan for the new city create a space structured by and for the intelligentsia. Indeed, Wren also applied his vision of experimental intelligence to the mundane and quotidian as he meticulously structured even the functioning of his own offices with a careful eye for administration; for the division of tasks between draftsmen, lawyers, engravers, and others; and finally for selling his plans to the politicians. As ambitious as his plan was, however, the reconstruction proposal that prevailed largely followed the prefire disposition of streets and properties in an attempt to minimize disputes and recommence the city's social and business life quickly. His vision for the water network, however, bore some fruit many years later.[26]

The exact date when the company asked for his recommendations for fixing its water distribution problems is not known, but it was around 1700.[27] Wren's report was later printed in the *Gentleman's Magazine*. In the report, Wren made clear that the ideal solution was not to apply a fix for the immediate situation surrounding the Soho Square supply problem. After investigating the matter, he declared that "I found myself unequal to give a pertinent opinion how it might be meliorated." For Wren, the only way forward was to rethink the entire distribution network as a complete system. The problems originated because the network was never built this way from the very beginning, "for the mistakes are fundamental in laying down the contrivances, and every day since new errors have been added, which are now inveterate." To make clear the importance of thinking of the network as an entire entity, Wren compared it to a diseased body. Rather than merely treating one part of it with palliative remedies, he would describe what a truly healthy body would look like, something he apparently thought the New River did not know: "But as a physician being called for to amend the distempers in a morbid body, must know what is naturally the anatomy and constitution of a sound body, I shall crave leave, as a naturalist, and somewhat versed in geometry, to begin from the first projection of this work, and consider as if there were not yet one pipe laid, what methods should have been taken from the beginning for the best and most equal distribution of the water brought to Islington; and when the right way is known, it is easy to judge and amend what is wrong." For Wren, the network needed to be conceived as a whole, and designed as such, using mathematics. The company's piecemeal design methods were no longer appropriate.[28]

Wren proceeded to outline in a series of steps how to do an integrated design. He recommended that the company be more careful to survey the land it was supplying to determine elevations better. Once this had been established, he suggested dividing the area between the New River Head and the Thames into four zones of approximately equal elevation and arranging the network to ensure that these areas all received approximately the same quantity of water. To do this, he suggested that the cocks or valves in the lower areas be of smaller diameter that those in higher areas, as the pressure was greater there, leading to high flow rates, while higher areas received only weak supply. Varying the valve size according to its disposition within the whole network would solve this, Wren argued. Further adjustments to valve size would need to be made according to the length of pipes serving a given house because the longer the pipe was, the slower the water flowed through it. The relation could be determined experimentally and recorded in tables. Larger valves would be needed toward the end of long pipes. Even this sort of adjusting for resistance to flow by changing valve size was not adequate given the actual length of pipes in use, and so Wren suggested installing small secondary cisterns for Soho Square and other more remote areas "to begin again the several branches that should be divided to the neighbouring parts." These cisterns should be fed by a large-bore pipe, perhaps seven inches in diameter, and then secondary pipes could branch out from the cistern. This ideal system should have been followed from the beginning, "but as all cities are built by time and chance, and not by mathematical designs; so the distribution of this noble aqueduct hath fallen by chance into the present œconomy, which altho' it serves pretty well the turn, is capable of improvement in time; and the more, by how much nearer it can be brought to such a primitive state, as is here rather wished for than advised; for an inveterate evil is often better tolerated than changed: yet much may be done by time."

Wren then enumerated what he thought the many errors the New River

Company had committed over the course of its chaotic history. He observed that the company had, in its earlier desperation for finding customers, laid mains with no regard to elevation and what effect this might have on water supply to the surrounding region. This meant far too much money had been spent on supplying too few homes, and the pipe network was in disarray. Some of the branch pipes were too long or were now blocked by air or filth. The lower supply regions were getting too much water, while those farther up were getting too little. Wren also rued that the small leaden supply pipes, which were attached to the company's wooden pipes, were all the same size, regardless of elevation. In addition, customers were not controlled in any way and could use water as they pleased, often leading to great wastage. The company, Wren claimed, needed to control this, perhaps through inspectors who had the power to enter houses, something the company could not do at present. He concluded his survey of the company's mistakes by stating that they could be fixed if his idea was followed: "They may, in a great measure be amended, if respect be had to the natural method before laid down." For Wren, the "natural method" of network construction was his overarching systemic and mathematical design.[29] Wren finished his report by considering whether building another reservoir at the New River Head at a higher elevation could solve the immediate problem (fig. 4.1). This may have referred to a proposal made by an artisan named Evan Jones, who in 1701 suggested building this reservoir and raising water from the lower pond into it using horse-powered pumps.[30] At this same time, the company also turned to John Lowthorp to make recommendations on how to address its problems.

John Lowthorp (ca. 1659–1724) was a clergyman, ordained a priest for the Church of England in 1683. After serving as a rector in Leicestershire, he lost his post when he became a nonjuror by refusing to take an oath recognizing William and Mary as the new monarchs following the deposition of James II. He moved to London and became the librarian to the Duke of Chandos, James Brydges. He was also a paid curator of experiments at the Royal Society, where he undertook experiments on the refraction of light in air, among other subjects.[31] Between 1704 and 1711, the New River directors consulted him as an outside expert as they had with Wren, seeking advice on how to reform the company's expanding network. In his first report of May 1704 to the company, Lowthorp considered the possibility of building a new reservoir at the New River Head up a hill some height above the existing reservoirs. The new reservoir could be supplied by means of a pump powered by wind, horse, or water. Although horses were likely the best, Lowthorp, like Wren,

Figure 4.1. *A view of part of the new River Head and new Tunbridge Wells at Islington* (1730–31), by Bernard Lens III

pointed out that the company's problems were systemic and could not be solved only through localized solutions, such as adding a new reservoir. To begin with, he pointed out that the company did not even possess basic information about its own works. The first step, therefore, was for the company to tally all its mains, recording their lengths, levels, and bores, as well as making a complete list of all its tenants and recording what they are charged.[32] To determine whether the supply problems were caused by inadequate water reaching the reservoirs at the New River Head, the company needed to know how much water the aqueduct could supply. Perhaps prompted by his own background as a Royal Society experimentalist, Lowthorp answered the question himself in a report presented in June 1704. He made experiments to determine the flow rate of the river. Although the flow was irregular, he deduced from observations made at one of the river's elevated aqueducts that it supplied at least 40,000 tuns of water per day (38.2 ML). Of this, about 17,000 tuns (16.2 ML) came from the River Lea, with the rest from the two original springs, mostly Chadwell. Lowthorp judged that this was sufficient to supply eighty thousand houses, more than all the houses in London and Westminster, and therefore the company's supply problems lay with actually distributing the water.[33]

The following year, Lowthorp was once again asked to make more recommendations because, in his words, "the complaints of ill service which the company so often receive both from their tenants and servants, make an enquiry into the true occasion of them allwayes seasonable but more especially at this time, when many of the members seem resolv'd to apply a compleat remedy however chargeable it may prove to them." Lowthorp once again stated that the problem was not one of volume of supply. Even the Soho Square supply problem that had so vexed the company was not simply one of sufficient water at the New River Head and the topology of the land: Soho Square was around ten feet below the level of the reservoir. The problem was "the unequall and unreasonable distribution of the water," with some areas, such as the West End to the north of St. James's Park, getting water for less than four hours a week, while the lower parts of the City had a thirty-six-hour supply. He did not think that adding a higher reservoir would be sufficient to resolve the problem. Lowthorp also recommended that the company redo its schedule for turning of valves to rebalance the distribution. Finally, he suggested that houses in the City where the elevation was the lowest be served one flight of stairs up from ground level rather than in the basement to bring their faucets closer to the same level as the poorly served ones in the West End.[34] This kind of supply to different elevations of houses had already been implemented in some parts of the city.[35]

In a couple of subsequent reports, Lowthorp indicated that the optimal solution lay with changing how the company's works were being managed, suggesting a centralized and integrated approach to management as well. He recommended hiring a "general surveyor" to be in charge of the company's overall operations, something that was not the case at that point.[36] Lowthorp claimed that supervision was divided between many people, and someone needed to be in charge to assimilate and collect information. Furthermore, the company clerk Ephraim Green did not have the skill or experience to estimate how much load mains and service pipes could bear. Lowthorp pointed out that no "mathematician" had yet figured a way to do this calculation.[37] The surveyor could, however, consult with the company's collectors and other workers to determine where new mains should be laid, together with their lengths and bores.

Reform of Operations

With all the discussion about reform and the proper way to design an integrated water network, the New River Company implemented a series of changes in the short term, as well as adopting other practices over the course

of the eighteenth century. By 1800, it had learned and settled into a sophisticated mode of operation that was stable and quite successful, enabling it
to grow and generate ever-increasing profits. Running a network of this size
required a strategy with many different parts. Because of the incompleteness
of the company's records, it is not always possible to determine when the
different elements of its overall methodology were developed. Some of these,
such as the use of valves to control water flow, were clearly in place from
the beginning, and their use expanded over time. Others, such as network
segmentation, appear to be new from the time of Wren's and Lowthorp's
reports, as no reference to them exists from an earlier period. Moreover,
Wren's and Lowthorp's comments indicated that there was no clear sense
of integrated design. Rather, the network had grown on an ad hoc basis. One
of the clearest recommendations coming from this time was the need for a
change in management. Lowthorp had suggested creating the post of general surveyor, and this was indeed done in the 1710s. Up until that point in
time, the company did not always manage its own infrastructure directly.
Not long after the construction of the aqueduct and the first pipes in the
city had been completed, the company farmed out the ongoing maintenance
and operations of its infrastructure to Hugh Myddelton for twenty-one years
beginning in 1622 for eight hundred pounds a year, a figure raised to one
thousand pounds in 1623 because of the losses he had incurred at the lower
level.[38] After Myddelton died in 1631, his widow Elizabeth was given a contract to perform the same work, but this soon descended into bickering over
her expenses.[39] In 1633 William Myddelton, Hugh's son, was given the contract for five years, renewed for twenty-one years in 1638.[40] William died in
1651, but the basic contractual arrangement was passed on to John Greene
(William Myddelton's son-in-law) and Gregory Hardwick in the 1665 for one
thousand pounds per year, down from the seventeen hundred pounds it had
reached before they got it.[41] Greene was then appointed the company's secretary, a position he held until his death in 1705.[42]

This mode of operating by which a company contracted its operations internally to people who were employees (or "servants") or otherwise insiders
was not unusual before the nineteenth century. The management of larger
enterprises was often handled by treating the firm as an agglomeration of
smaller ones, through subcontracting within the firm itself rather than by
developing internal hierarchies.[43] The New River began to deviate from this
practice when it hired Henry Mill as the surveyor in the 1710s. From this

point forward, subcontracting on the basis of a fixed contract ended, and the company ran most of its own operations through the surveyor and the collectors. The directors oversaw the operations in detail in its weekly meetings. Little is known about Mill, the person who took on this new role, but he held the surveyor's post until 1767. Although the title of surveyor had been used before he held it, it was only with Mill that the new role of what later came to be the chief engineer was created. He was referred to as "engineer and surveyor of the works" from at latest 1725.[44] He and his successors assumed the responsibility that Wren and Lowthorp had recommended, that of oversight for all technical matters, and he was answerable directly to the directors. The New River's second surveyor was Robert Mylne (1733–1811). Mylne was born in Edinburgh from a long line of masons. He apprenticed to a local carpenter, and later went to France and Italy with his brother to study architecture, including with Giovanni Battista Piranesi, who encouraged him to pursue his interest in the water systems of ancient Rome. He won an architectural prize in Italy, which helped to secure his first work once back in Britain: the new Blackfriars Bridge in 1760. It was complete in 1769. He thereafter held posts such as surveyor at St. Paul's Cathedral and, in 1767, assistant surveyor at the New River Company. He was also elected to the Royal Society that year. He became the company's chief surveyor in 1771, and helped John Smeaton found the Society of Civil Engineers. He somewhat overextended himself, becoming involved in land management, canal engineering, and architectural work, some of which, such as the Gloucester and Berkeley Canal, ended in failure.[45]

Besides adding the key post of surveyor and chief engineer, the company developed a methodology during the eighteenth century that can be broken into five parts: maintaining adequate reserves at the New River Head reservoirs; design of the pipe network; load management by control of users; pipe manufacturing; and, finally, repair and maintenance of the network. Some of these elements bear quite clearly on the company's core operation of distributing water, but others, notably pipe manufacture and the repair and maintenance of the network, were essential supporting activities. The support activities were vast operations, and they consumed more resources, from the point of view of money spent and employee time, than anything else the company did, including water distribution. Indeed, the New River Company was also in effect a manufacturing company, with the making and laying of pipe as one of its primary activities. The job of distributing water effectively implied a commercial and manufacturing operation to purchase and process enormous quantities of wood every year.

Maintaining Adequate Supply

As Lowthorp and Wren both realized, the first step in distributing water well was having enough water at the reservoirs of the New River Head. This in turn meant drawing enough water from the sources, getting it to the Islington, and keeping a sufficient buffer there for distribution. Drawing enough water from the New River's sources was a political problem, not a technical one. Its original acts gave it clear rights to the water from the two springs, and these were never controversial. The third source, the River Lea, was a different matter entirely. It was the only one among the New River's sources that had any possible excess capacity, and as the New River grew, it naturally sought to draw more from that source. Other parties, however, notably mill owners and barge operators, had interests in the Lea's water, producing many disputes with the New River. The first in the 1660s had forced the New River to reduce the size of the pipes it was using to take water from the Lea from one its channels, called the Manifold Ditch. Bargemen and mill owners were not satisfied with this and soon complained again to the City and the Crown about the New River's greed. In 1672 Wren in his position of surveyor general exonerated the company from the charge after an investigation.[46]

The dispute flared again in the 1730s because the New River was able to increase how much water came through the two pipes by judiciously expanding the Manifold Ditch and building a dam or sluice across part of it.[47] This ensured a high head of water over the pipe intakes and therefore greater flow. The company also installed a "balance engine" to control how much water was being drawn. The engine was a wooden float on the Manifold Ditch designed to keep the flow into the New River constant by regulating the sluice gate feeding a trench. It also installed a gauge in the trench. The gauge was a trough, six feet wide and two deep, used to measure water flow and to ensure it stayed at the allowed maximum indicated by the gauge. If the flow through the gauge was too high, the New River itself would overflow downstream into surrounding fields.[48] The 1730s edition of the Lea disputes was resolved by agreement with the trustees for the River Lea navigation in 1736. The agreement required approval from Parliament, which was received in 1739, although the enabling law was fiercely opposed by Lea bargemen and mill owners. The law allowed the New River to draw water through the gauge for £350 per year paid to the trustees.[49] The dispute erupted again in the 1780s and 1790s, but the New River continued to take most of its water from the Lea.[50]

The company also had to maintain the aqueduct itself in good shape to keep the water flowing along its gentle slope, a task requiring full-time employees. The entire length of the river was divided into eight or so walks patrolled by a walksman whose job was to keep the river clear, remove obstacles fallen into the river, and ensure the banks were whole. The banks sometimes gave way, causing flooding.[51] People occasionally tried to steal water by cutting the banks of the river. This was a sufficiently present concern that the company repeatedly and successfully petitioned the Crown for proclamations enjoining people not to damage the banks.[52] The walksmen also had to cut the weeds out the river because, if they grew thick enough, they could clog the river and noticeably decrease its flow.[53] Walksmen could be fired for neglecting this.[54] Another regular problem was mud, and laborers had to go into the river regularly to dig it out, sometimes dumping it on the surrounding land, much to the neighbors' consternation.[55] If the mud was allowed to accumulate in the river, it would widen and grow shallower, slowing the flow and raising the risk of its freezing in the winter.[56] When that happened, as it did in 1789, the river could overflow and flood surrounding fields.[57] In another case in 1795, water supply stopped entirely in the whole city when the river froze. Laborers were sent out to break up the ice.[58] Another means of controlling the water flow on the river was by sluices. Where the water would otherwise run too rapidly, the company installed sluice gates to pen it in and slow the flow. These sluices were also a means of controlling flooding by creating defined spaces where excess water could accumulate.[59] There were forty-three sluices on the river.[60]

Finally, having adequate supply also meant maintaining the reservoirs at the New River Head. This was largely a question of capacity, something the company addressed by constantly adding to and expanding them over the years. Indeed, one of the first outcomes from all the early eighteenth-century reports and investigations regarded supply. To improve supply, around 1707 the company hired George Sorocold (ca. 1668–1738), an engineer who had extensive experience building waterworks throughout the country. Before coming to the New River, Sorocold had improved or built waterworks in many provincial towns, including Derby, Bristol, Leeds, Newcastle, Norwich, Portsmouth, and Sheffield. He was also involved in river improvement schemes, such as on the Aire and Cam. In 1693 a patent had been award to him and

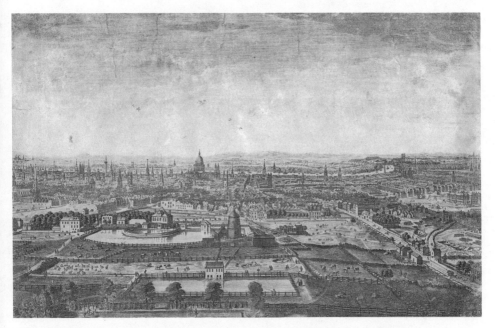

Figure 4.2. The upper pond is in the foreground, with the windmill tower and lower ponds behind. Thomas Bowles, *North Prospect of London taken from the Bowling Green at Islington* (1753), original from Canaletto (ca. 1745)

John Hadley for a waterwheel that rose with the water level. Sorocold and Hadley had already worked in London in 1696 for Marchant's works, where they designed their pumps.[61] As discussed in chapter 5, Sorocold also did important work in rebuilding the London Bridge Waterworks after 1700.

For the New River, Sorocold designed the new higher reservoir, or the upper pond (fig. 4.2), that had been recommended and investigated over the previous few years.[62] The pond was finally built in 1708 with an elevation 33 feet (10 m) above the original reservoirs.[63] Water was pumped into this new reservoir at first by a windmill with an uncommon design, having six sails. By 1726, the windmill had been replaced by horse power because of twice sustaining damage in windstorms. It was not repaired after the second storm, and although the directors considered bringing it back into operation in 1742 when more water was needed in the West End, it remained derelict.[64] Greater pumping capacity came with a steam engine that was installed first by John Smeaton in 1768. In order to hold the greater volume that the steam pump raised, the upper pond's capacity was increased in 1780 by raising its top height another 18 inches (0.45 m).[65] Finally, the Smeaton steam engine was replaced by one supplied by Boulton & Watt in 1786.[66] The pumping to the upper pond meant that it required more active care than the lower ones, which filled from the New River directly. If the pumping was slow and the level in the upper pond fell too low, the mains supplied from there would not be full, leading to bitter complaining. One notable case in 1792 saw the level fall 23 inches (0.58 m) below capacity and supplies in some parts of town grow very weak.[67]

There were a few other minor issues with maintaining reservoir capacity, such as silt. Like the aqueduct itself, the company had to dredge mud out of the reservoirs, although it did so less frequently. The inner pond was cleaned in 1723, but not again until 1774 when seven thousand loads of mud were removed.[68] Finally, excess water from the reservoirs drained through a waste sluice into a small pond and then to the River Fleet that ran close by. A waterwheel was set up to use this waste flow as it entered the pond. Being held in a smaller reservoir, this water was dirtier, and so when the company decided to use it to serve the Clerkenwell area, there were many complaints about slow flow and the filth of the water.[69]

Steam engines had been introduced to the London water industry many years before the New River adopted one. Although they were to prove important for the water industry in the late eighteenth century, they had a long difficult history before they become reliable. They were initially adopted by the York Buildings Company not long after Thomas Savery first patented a design in 1698. He seemed to have persuaded the owners of the company to install one around 1713, but it caused no end of troubles. Prone to breaking down, it also required its valves to be actuated manually. It belched thick black smoke to the intense annoyance of the neighborhood and was eventually "look'd upon as a useless piece of work, and rejected." Meanwhile, Thomas Newcomen designed his improved atmospheric engine and came to an arrangement with the owner of the Savery patent, who had died in the meantime, to sell his engine under that patent. The York Buildings Company tried again in 1725, erecting the first Newcomen engine in London to pump water to a reservoir in Marylebone via a water tower set up along the Thames. This one remained in operation longer than the first, but it too was abandoned, in 1731, because it simply cost too much to supply it with coal. The company, furthermore, was in financial difficulties, and sold its Marylebone reservoir to the Chelsea Waterworks.[70] The first engines that were successful were two Newcomen engines set up by the Chelsea Company in 1741–42 (fig. 4.3). The Chelsea Waterworks had faced severe supply

Figure 4.3. The Chelsea Waterworks' Boulton & Watt steam engine. *A View of the Fire Engine at Chelsea Waterworks* (1783)

problems for some years previous and was even losing tenants embittered by the poor service. Its pumps were originally driven by tidal mills situated on the Thames. Water was allowed to flow into reservoirs at high tide. Some was let back out as the tide ebbed, turning a waterwheel pump that sent water to higher reservoirs located in St. James's Park and High Park. The steam engines solved the supply problems, and most of the grumbling abated thereafter. For its part, the York Buildings Company tried again in 1752 with another Newcomen engine, and it remained in use for many years. The Shadwell Waterworks also installed a steam engine around 1750.[71]

Although steam engines were part of London water system supply from the 1710s, their importance before the 1760s should not be exaggerated. The New River did not have one, and the LBWW used its steam engine, erected around 1760, only intermittently. These two companies dominated supply until the late eighteenth century. The other steam engines belonged to companies that depended on them, but with the exception of the Chelsea Waterworks, they were all small companies (see chapters 5 and 6 for comparisons). The early steam engine models were unreliable and too expensive to operate, but by the 1770s more reliable and efficient designs, including those prepared by

Smeaton, were introduced. By 1775, there were ten steam engines operating in London for water companies: two each at the York Buildings Company, the Chelsea Waterworks, and the Shadwell Waterworks, with the New River, West Ham Waterworks, LBWW, and Lambeth Waterworks possessing one each. There were still fourteen waterwheels in operation at that time, most owned by the LBWW. John Farey estimated that the combined power of the waterwheels and the steam engines was about equal at 105 hp (78 kW).[72] As important as the waterwheel would remain for industrial mills into the nineteenth century, its era was slowly coming to a close in the London water industry.

From the 1780s onward, water companies relied almost exclusively on steam engines for new pumping capacity, although existing waterwheels were maintained for decades.[73] Boulton & Watt had made further improvements to the steam engine with a separate condenser and sold its first one in London in 1776 to a small distillery, while the Shadwell Waterworks made the first purchase by a water company in 1778. Another followed first for the Chelsea Waterworks in 1779, then another for the Shadwell Waterworks in 1784, and finally for the LBWW and the New River. Some were also sold to companies on the south bank, including the Lambeth Waterworks and the Borough Waterworks.[74] Boulton & Watt further improved the efficiency of their steam engine by introducing the double-acting piston, where power was developed in both directions of the piston stroke, rather than in only one direction as with the earlier models. The first was sold to the New River in 1794, and many others followed. By 1800, when the original patent expired, Boulton & Watt had sold ten of its steam engines to London water companies.[75] The largest engine the firm sold was in 1804 to the Chelsea to replace its original 1741 engine. It was made almost entirely of iron and developed about 43 hp (32 kW).[76] The Chelsea added two more in 1809 and 1818.[77] Many new companies were founded in London after 1806, and all of these used large steam engines for pumping, including the East London Waterworks, the West Middlesex Waterworks, and the Grand Junction Waterworks.[78] They grew rapidly, taking on tens of thousands of customers within a few years. The New River continued to use gravity for much of its supply, but by this point, most of the rest had shifted to steam, and even the New River used steam to pump to the West End. On the basis of calculations of market share and estimations that about 10 percent of the New River's water flowed through its upper reservoir, which was filled by pumps, steam engines went from providing water to about 15 percent of houses receiving water in 1750, to 20 percent in 1775, to 35 percent in 1800, to 60 percent in 1820.[79] It had come to dominate supply, and the London water industry's rapid expansion in the early nineteenth century relied on constant improvement in the efficiency of the steam engine. With the removal of London Bridge, all water-wheels were gone. Only the New River's gravity-powered supply from its perch atop Islington remained not relying on steam.

The Pipe Network

The pipe network itself was the largest part of the New River Company's infrastructure and the most complex to maintain and control. Its role was, evidently, to distribute water as reliably as possible according to the company's ideas about which area of London should receive how much. The company used various strategies to achieve this, although adjustments were constantly needed, given the constantly changing shape of demand in the dynamic metropolis. One of the techniques the company used to improve reliability was a system of dual pipes that was implemented over the course of the eighteenth century (fig. 4.4). Larger pipes, called *mains*, were used to transport water from the New River Head to various districts of the city. Smaller pipes, called *service pipes* or simply *services*, were attached to these mains and ran through the street to distribute water to homes or public courts. The connections to buildings were lead pipes called *ferrils* that linked into the wooden services (fig. 4.5). As the eighteenth century progressed, the company tried to differentiate between the mains and services more clearly. This meant that it avoided having ferrils drawing water from the mains. To further isolate the ferrils from the mains, the company began using another set of pipes called *riders* that ran beside the mains to provide water service to the buildings on those streets, thereby removing the need to connect them directly to the mains.[80] This system ensured that water flow in the mains remained relatively constant, even as it was distributed across various zones through service pipes. A customer who let water run to waste extravagantly would have less an impact on water supply in the surrounding area because he was more isolated in this dual system. It also served as a means of isolating service areas into zones of approximately equal elevation, fixing the problem that Wren and Lowthorp had identified in different ways. Eventually all the principal mains had riders associated with them.[81]

Within the New River's network, the shift to the double arrangement of pipes was quite slow, perhaps a reflection of the chaos in the pipe network and the difficulty in reforming such a widespread system that had grown

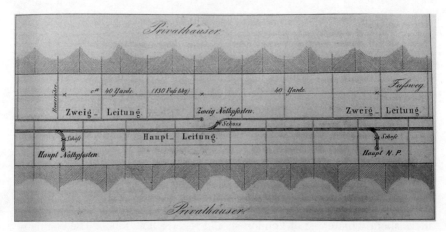

Figure 4.4. The doubled pipe system from 1844. The main (Haupt-Leitung) feeds service pipes (Zweig-Leitung). All houses are connected only to service pipes. William Lindley and William Mylne, *Ingenieur-Bericht, die Anlage der Oeffentlichen Wasser-Kunst für die Stadt Hamburg betreffend* (1844)

over the course of a century. Although some distinctions between mains and service pipes can be found in the company's records from the early eighteenth century, the directors ran a couple of concerted campaigns to move all customers off the mains and onto service pipes in the 1770s and 1780s (fig. 4.6). In late 1782, the directors constituted a special committee to examine how to transfer tenants off mains onto services.[82] Only some brewers were allowed to continue drawing from the mains.[83] Besides the campaigns, references abound in the company's minutes from this period, ordering the installation of new service pipes in streets so as to move any customers in those streets from the mains to service pipes. Customers, however, were not so keen to see this change implemented. For them, being connected to a main meant a much better water supply because it was at higher pressure and flowed more regularly. Nevertheless, the company maintained a strict rule that no customer should have any connection to the mains, with the exception of a few cases where no service pipes existed. The company tried to keep track of where these were located.[84]

Another strategy the company tried to use was to segment its network to better isolate higher supply zones from lower ones, another of Wren's suggestions. If the company's employees noticed that one area was draining water from another, the directors would order a valve to be installed to shut water back in the higher zone.[85] In 1787, for example, a valve was placed on

Figure 4.5. Wood water pipes at Clerkenwell around 1815. Alfred Stanley Foord, *Springs, Streams and Spas of London* (1910), p. 296

the Berkley Square main in the West End to isolate the portion that served higher areas when it was feeding the lower ones.[86] The company also tried to ensure more even flow throughout parts of the network by building interconnections between the mains. There were, for example, a number of mains running to Brick Lane in the east end of the city, and in 1787 connections were installed between them to even supply.[87] Similar orders can be found throughout the company's minutes, such as for links between the old and new Goodman's mains, or between the two branches of the New Road main in Portman Square.[88] Furthermore, the company tried to build secondary cisterns as Wren had recommended, although whether his recommendation was the instigation for the idea is doubtful. Wren's specific suggestion for Soho Square, for example, does not seem to have been built. In general, adding cisterns was not easy because they required more space than was available in most cases. The most notable example was one installed at the north end of Tottenham Court Road, once again to serve the West End. The company purchased the land for this in 1790 and built a dedicated main to keep it full. A number of the West End mains were then supplied out of this new reservoir.[89]

This system of segmentation into zones of supply of different elevations,

Figure 4.6. The New River's pipe network in 1770. Reconstructed from Collectors' rents and arrears book LMA NR ACC/2558/NR/12/001 and overlaid on Bowles's *Reduced New Pocket Plan of the Cities of London and Westminster with the Borough of Southwark* (1795)

with a clear distinction between mains and service pipes, together with interconnections to make the network more robust, was sufficiently established that when in 1735 the architect Richard Castle (also Cassels) (1690–1751) wrote a pamphlet on Dublin's water supply, he explicitly recommended the main-and-service-pipe model the New River used. Castle had lived in London around 1725–28, where he became interested in waterworks. He then moved to Ireland and was consulted by the Dublin Corporation's water committee on improving the city's water supply. In his report, he described how in London mains ran through streets without connecting to buildings, and then service pipes would double back to reach those same buildings the mains passed. He also claimed that the New River, which he thought offered the best model to emulate, had secondary cisterns to ensure "the more equal distribution" of the water. In addition, he explained how the system of ball-cocks in houses prevented the overuse of water, something the New River had inaugurated.[90] In similar vein, John Smeaton in 1761 wrote a report for the Halifax City Council about its waterworks in which he recommended using the main-and-service-pipe system with network segmentation to control for difference in elevation.[91]

Besides the double system of pipes and network segmentation, the company also tried to ease flow by removing components that caused excess friction. There had always been an intuitive understanding that obstacles or roughness in pipes reduced flow, and so the company occasionally made decisions specifically to reduce friction. For example, in 1794 a brass valve slowing flow in the Oxford and Soho mains was removed. Another frequent change involved moving tenants from long service pipes to shorter ones where this was possible because flow along long pipes was poor.[92] Whatever changes an empirical approach to friction may have suggested, the mathematics to describe what sorts of obstacle caused greater or lesser friction was increasingly explored in the late eighteenth century. In the 1770s and 1780s, the French military engineer Louis Gabriel Dubuat and the priest Charles Bossut both published treatises on hydrodynamics in which they described the effects of pipe features, particularly elbows, on water flow. Dubuat, who was Bossut's assistant for some of his work, made experiments of his own that he claimed revealed how the resistance to flow increased with the sine of the angle of the elbow turn. He argued that angles greater than forty degrees in particular greatly reduced flow. Bossut also devoted extensive space to mathematical descriptions of the effects of pipe size and elbows on flow.[93] Their work was soon available in Britain in English translation. The Scot-

tish natural philosopher and inventor John Robison in his article on "waterworks" for the 1797 edition of the *Encyclopædia Britannica* expounded on
the usefulness of Bossut's and Dubuat's work for the analysis of pipe flows.[94]
To what extent their specific contributions had an effect on the New River is
unclear. Within the company's minutes, there is no clear indication that any
analytic means were used to decide optimal pipe configurations. In addition,
water companies had recommended in the 1740s that right-angle elbows be
replaced with curved ones.[95]

Nevertheless, there was a trend from the 1790s showing an increasing appreciation for how important friction losses caused by small and rough pipes
could be. From that time, the directors gave many orders for changes to configurations that made specific reference to easing flows in a way that was not
in evidence earlier. In 1790, for example, cross joints that included right-angle
turns were removed from service pipes and replaced with Y-shaped joints
called three-hole pieces because these featured lesser angles.[96] Another decision to reduce friction was made in 1792 when the directors resolved to stop
using two-inch pipes, making three-inch service pipes the smallest ones in
use, the ferrils excepted.[97] In 1798 the company, realizing that a sharp turn
was inhibiting flow, re-laid the Dover main from Compton to Brewer Street
along "a more regular sweep to avoid angles."[98] In a similar vein in 1802, the
company agreed with a householder to change the shape of his wall on a corner of Coppice Row Clerkenwell from square to rounded. This would allow
all the mains at that corner to be reset into a round shape, greatly improving
the water supply to the West End.[99] These sorts of decisions may have reflected some influence of mathematical analysis of the type Dubuat espoused
on network design. Even if adequate mathematical techniques to solve for
flows in pipes and especially networks would not come until the late nineteenth century, mathematical fluid analysis helped water engineers to make
decisions by showing how high the friction through tight corners could be.
Although in general it is difficult to discern any influence of contemporary
natural philosophy on the water industry, this particular case of fluid dynamics for pipe design may be an exception. Dubuat and Bossut had practical concerns. They wanted to produce results that could help the design
of ships, canals, aqueducts, and pipes. Their results may have had practical
consequences for the New River at least.

One of the more important strategies the New River used to maintain adequate supply was load management. That is, the company tried to ensure
that the supply capacity of pipes, mains, and entire districts was adequate

to the local demand for water. Although the term *load management* was not used, the practice was evident in the company's actions. Due to the organic growth and constantly shifting number and type of tenants, load management was an ongoing task. What had worked adequately for some time could result in degraded supply if too many tenants had been added to a main or service pipe. This happened particularly in rapidly growing areas, which in the eighteenth century meant mostly the West End. In order to match load to capacity, the New River used various techniques at different levels corresponding to pipes, mains, and districts. The simplest techniques worked at the lowest level, which was adding service pipes. If the supply on a pipe was poor, the workers could cut a service into two sections and attach both to the main, effectively doubling the amount of water served.[100] Another option the company had to improve local supplies was moving tenants around to different service pipes, rearranging the pipes themselves, and installing new ones. The company could also install another valve at a certain point along the pipe. This valve would then be added to the turning schedule, which would shut the water flow back and force more supply into buildings upstream from the valve.[101]

The next level up in load management involved transferring pipes between mains. This took place, for example, in the Portland Square area in 1787, along with some of the lower-level options. The company transferred the pipes in Paradise Street, Paddington Street, and Nottingham Street from one main to the north branch of the New Road main. The service pipe in Portman Square was shortened and attached to end of the same main, as was the Baker Street service. The pipe in George Street was cut in two with valves so that the east and west ends were served separately. A pipe was then laid from George Street up Manchester Street to take on more tenants.[102]

At an even higher level of load management, the company could add an entirely new main. In 1787, for example, the service pipes close to the Thames supplying Water Lane, as well as White Friars through Silver Street and Lombard Street, were taken from the south side Fleet Street service and transferred to a new main running in Fleet Street between Fetter Lane and Fleet Market, which had been installed to serve Hatton Garden.[103] In another case, the tenants connected to the four mains then coming from the upper pond were not being well served due to the many new buildings being added around Baker Street. The company explored how best to bring "relief" to these mains and decided to lengthen another one into the area. They added 1,400 yards (1,280 m) to the north branch of the New Street main, bringing it as far as Portman Square.[104] The number of hours of service provided into those mains was also increased from 283 hours per week to 301.[105] Because the mains were not always full of water, adjusting the number of hours they were supplied was another high-level load management tactic, equivalent to adding a new main from the point of view of getting more supply to a certain zone. The company had recourse to this tactic regularly.[106] In the case just mentioned, adding an extension to the New Road main was not the end of the matter. In 1789 service to many parts of the West End was deemed once again to be poor, and the board commissioned a report "to give a more permanent supply of water to the West End."[107] The result was that the duration of water supply service was lengthened on many West End mains, such as two more hours per day to the New Road main, and one hour per day to the Oxford and Portland mains. The directors, however, judged that the New Road main was overloaded and ordered that an entire new one to be driven to the street. It was 1,340 yards (1,225 m) long and cost £630.[108] Even this solution proved inadequate, and upon further study the next year, the company installed a sixteen-inch iron main from the high pond, 2,215 yards (2,025 m) in length and costing £6,976.[109] Although the water companies had used iron pipes before this time for shorter lengths, such as over sewers, this was the first occasion the New River used the material for a long main. Many of the West End mains were then connected to the iron main, including the Bedford, New Road, Cavendish, and Portland mains.[110]

A further, top-level option to match loads with supply capacity was to shift entire mains between reservoirs.[111] In one case in 1788, the New Road main was moved to the high pond. However, some areas it served, notably around Tottenham Court Road, did not need this greater pressure, and a new main was laid from the low pond to take them on.[112] In another example from 1788, an interconnection 160 yards (146 m) long was installed not far from the New River Head between the Soho main coming from the high pond to the Cavendish main serving low pond water. The higher pressure from the upper pond allowed all the areas north of Bedford Square to be transferred from the Soho main to the Cavendish main, leaving more capacity to the Soho main to supply other areas.[113] Later reorganizations made this unnecessary, and the connection was removed in 1794.[114]

Load management was a strategic issue and involved physically amending the water infrastructure. By contrast, the process of actually delivering the water to customers was a daily and hourly operation that the load manage-

Figure 4.7. A turncock. Drawing by H. W. Petherick. *Aunt Louisa's Welcome Gift; London Characters* (1885?), p. 3

ment was meant to enable. Because of the scheduled nature of water supply, the New River's network was quite dynamic, with water pressure and flows changing hourly across pipes and mains. Water was constantly shifting between different zones and pipes over the course of a day, and different days of the week had different schedules. This meant that the process of distributing the water was also labor intensive. It required the turncocks to develop a routine for turning all the valves in the streets to send the water to the right sections and pipes (fig. 4.7). Each turncock had his own walk that he traversed from early in the morning, often around 4 AM, opening and shutting valves with a key he had to guard very carefully.[115] If the key was lost, as happened in 1787 (the misplaced key was eventually found with a pawnbroker), the valves needed to be replaced.[116] There was in addition a chief turncock at the New River Head who controlled the flow from the reservoirs into the mains over the course of a day. As the walks grew to span a few hundred streets, the set of turnings along them became complicated. Mistiming a turning or neglecting one would result in complaints of poor service. In 1786, for example, the turncock William Becket neglected his turnings on the high-pond mains going to Islington and New Road, leaving "the water to be longer on for some hours [more] than allowed to the great detriment of this company." He was given a "severe reprimand" and warned that a recurrence would lead to his discharge.[117] The knowledge of the location of the valves and the turning schedule lay with the turncock, but the company did not want to be at his mercy should a replacement be needed on short notice. It compiled and kept books of turnings, codifying and recording the turncocks' tacit knowledge. The company placed great importance on these books because they represented the written versions of the quotidian routine keeping the distribution of water well ordered. The books were a way of minimizing the scope for error—and discretion—on the part of the turncocks. How highly the company valued these books was reflected in a 1787 order that "a neat sett of turnings on all the high pond services be established to be left in this office and a strict compliance therewith required on pain of expulsion."[118] In another episode in 1799, the directors reprimanded turncock John Powell for "not doing his turnings regularly," and he "was ordered not to deviate from the rules in his book of turnings."[119]

All these levels of load management, from the service pipes up to the reservoirs, in addition to the actual daily delivery as managed by turncocks, were centralized and controlled through maps and inventory lists of valves, main, and pipes. Both Wren and Lowthorp had raised the issue of mapping and inventories when they pointed out the company's lack of knowledge of its own network. Although it had tried to keep some sort of inventory before 1700, maps would go out of date quickly because mains, pipes, valves, interconnections, and splits were constantly being adjusted.[120] The company nevertheless tried to keep up with these changes. In 1742 the directors resolved to keep a new book listing all mains and riders, as well as a map showing the locations of all its pipes.[121] This practice was maintained in the late eighteenth century, and the company relied on having large maps of London at its head office showing all street mains and pipes, as well as an inventory of all pipes and valves in the streets, listing their sizes.[122] The books detailing the turncocks' turnings—as well as the collectors' walks—were part of this process of creating maps, inventories, and lists to control the functioning of the network.[123]

Controlling Customers

Another important element of managing the load on the network was controlling how tenants received water. Specifically, the company over time began using more means to control supply into buildings. Originally, the ferrils into homes had no valves, and water would flow into the reservoir after the valves along the mains and service pipes were opened by the turncocks. Even if there was a valve, some people would leave them open all the time, allowing water to overflow the cisterns, and from there the water would run into the streets (fig. 4.8).[124] Over time, the company thought that some of its customers were drawing far too much water and abusing their supply.[125] This problem of drawing too much water was not, however, limited to paying customers. Wren had mentioned that there were "interlopers . . . meddling with the company's pipes," presumably attaching ferrils to the company's pipes and mains without the company's knowledge.[126] Ways of addressing this included legal means. The company petitioned the Crown for special powers and edicts against such meddling, even as early as 1638. In a proclamation from 1669, Charles II enjoined people not to "intermeddle" with the company's pipes without its permission and urged "sherrifs, bayliffs, constables," and other officers to be vigilant to prevent this. In a new proclamation, James II granted the New River powers in 1686 to search houses in the presence of a constable if there was suspicion of abuse.[127]

The company took theft seriously because it was a constant problem. If someone was discovered to have been stealing through a surreptitious connection, the company would sue them for some years of back rent. The threat was serious enough that those accused would often pay.[128] This did not, however, eliminate the problem, and surreptitious connections were perennial. In 1784, more than a century after the first proclamation, the directors thought that "plumbers have frequently been employed to put water ferrils into the mains or service pipes . . . and to join the lead pipes out of one house into another . . . by which many persons have been supplied with water without paying any rent."[129] House owners would sometimes bribe the company's own employees to install unauthorized connections that remained off the company's books.[130] In 1792, for example, a collector reported that "Mr. Bailey the dyer in his walk hath lately been clandestinely laid on to the main" by a New River foreman whom he had bribed. The dyer had also bribed the turncock with a half guinea to operate this valve. The turncock was fined a guinea for his indiscretion.[131]

It was not only single exceptions that worried the company. In 1784 the directors became convinced of the existence of widespread theft and ordered sweeps to be made of the Whitechapel district. They discovered 525 houses that were not in the collecting book, and adding them increased the yearly rental by £280 from £402 initially, with another 500 houses still to be verified.[132] Continued inspection of the Whitechapel district added yet more houses, bringing the total discovered to 896 houses, representing £477 per year new rent.[133] In 1787, a new audit revealed another massive problem with theft in the Whitechapel and Fleet Street areas, seemingly caused by incompetence or corruption among the collectors. The company ended up charging £871 in fines to legitimize illicit connections to 1,089 houses, and thereby adding £560 per year to the rent rolls. The sweep brought in even more revenue than this because the company also raised fees for those already on the books who were using more than allowed. This increased the rents by a further £90 per year.[134]

Figure 4.8. An eighteenth-century lead cistern in 46 Berkeley Square, Westminster. Courtesy City of London

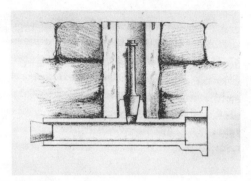

Figure 4.9. A fireplug. William Matthews, *Hydraulia* (1835), p. 309

Theft could also occur in other ways, such as in 1798 when a brass founder was discovered to have a key to one of the company's fireplugs, which was a valve located under the street to be used only in case of fires (fig. 4.9). Somehow, he had obtained a copy of the key and would regularly open the plug to feed his steam engine.[135] Theft from fireplugs happened often enough that the company printed and distributed notices that anyone caught stealing water through them would be prosecuted.[136] In another case from 1808, the New River surveyed the houses on the Minories by entering them because no one knew which houses were getting water, or even which company was supplying them. Once completed, the company determined that most were probably stealing connections from the LBWW and sent the list to that company to look into.[137]

In addition to threats of fines and inspections to control water use, the company also used technological innovations. One such measure was developed around the 1720s with the introduction of ballcocks (see fig. 4.10). These were valves regulated with floats that shut the water intake pipe once the water level in the reservoir reached a certain level, taking the regulation of water flow out of the hands of the buildings' occupants. In 1724 the company's directors ordered that all tenants must have such ballcocks or face being cut off.[138] The order was unevenly applied because the New River could not control how people built or modified their houses. It was, however, a general principle to which the company regularly had recourse, especially when it was having more difficulty in getting water to some areas.[139] It also inspected new houses to ensure that reservoirs and ballcocks were present before connecting them.[140] The problem with people letting water run to waste was by no means new. In 1594, soon after Bevis Bulmer first

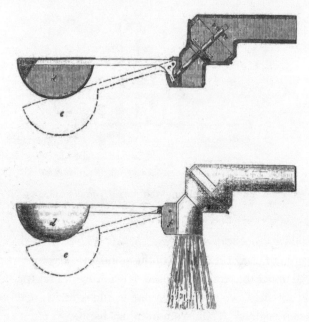

Figure 4.10. A ballcock. *Mechanics Magazine* 32 (1840), p. 457

established his waterworks, he reserved the right to enter people's houses to ensure that they were not letting their water run, and if this was deemed to be the case, then the householder would be fined two shillings and sixpence. If the problem persisted after a century, it is not likely he succeeded in eliminating the practice.[141]

Heavy users of water deserved special attention: not only were their fees very high but their demands could also have a big impact on the local network. Since no water meters were then in use, the value of a water supply contract depended on judging what sort of water-using activities would be carried out by the tenants. Large users—typically taverns, sugar refiners, distillers, brewers, stables, and, later in the eighteenth century, premises that used steam engines—would sign special contracts allowing them more water. Sometimes even a heavy user such as a distiller was discovered stealing water by a hidden ferril attached to the mains.[142] For this reason, the company tried to monitor how tenants were using water and to react to special cases where usage was high. If a building was using too much water, the water companies' collectors and foremen would try to inspect them to discover the cause.[143] Higher fees would be demanded if the usage had increased above what was expected. Sometimes, however, tenants could start using more without notifying the company and would continue paying the old rates.[144]

The company accommodated heavy users in different ways according to the circumstances. For the most lucrative, ways were found to meet their needs, by, for example, installing valves farther along the pipes to shut back the water for the user.[145] Another option if a particularly heavy user could not get enough water from the service pipes was to allow direct service from the main.[146] This did not mean that these customers were given carte blanche to use all the water they wanted. Generally, if these tenants did not have a ballcock, one would be installed, such as for a distillery in 1784.[147] Stables presented their own challenges because often they were served outdoors in a court via a fountain, and many people could easily take water from it.[148] The company tried to have the stable owners restrict access to these fountains. In other cases, if a large customer was deemed to be harming his neighbors' supply and was not paying enough, the company could even switch them back to a service pipe.[149]

The late eighteenth century saw the emergence of a new kind of heavy user of water: customers with steam engines. Steam engines had seen limited use in London in the early eighteenth century, but in the 1780s when Boulton & Watt's rotary engine became available, they found wider applications. Some of the New River's customers had them installed and used their water supply for the engines without informing the company, such as in 1790 when a tobacconist was found to be using New River water for its steam engine.[150] In 1790 a brewer in White Cross Street requested water for his steam engine to supplement well water during dry periods. Because the brewer was already using water well in excess of its lease, the company refused. The brewer, however, agreed to a new lease that specified changes in how he was using water, prompting the New River to agree to his proposal.[151] In another case from 1799, the brewer Sharpe in Golden Square was told that "he must so alter his works as to leave no possibility of using New River water for his steam engine—as [sic] if he does use it, the company will totally take away their water."[152]

The collectors were the primary interface between the company and its tenants. They were in effect the public face of the company because they met with all the company's customers several times a year, either in the customers' homes or in a designated coffee shop within their district. Because they managed customers and were supposed to ensure that all was well within their assigned areas, they had an important role in running the New River. As with the turncocks, the company codified in writing the standards of behavior it expected its collectors to meet. While customers could also complain to turncocks about poor service, the collectors were ultimately in charge of the district and gave orders to the turncocks and the foremen. The company divided London into districts, with each one assigned to a collector. In 1708, there were ten collectors, rising to fourteen by 1739, a figure it held until the 1790s when there were fifteen.[153] The company increasingly tied these collectors to their districts. From 1757 all collectors were required to reside within the metropolis, and from 1778 they had to live at the center of the district so as to be easily available.[154] They were even forbidden from leaving London for any time unless the directors at the weekly meeting at the head office had permitted it. The company expected them to know their districts intimately and supervise all aspects of company's operations there.

The collectors signed all leases with customers, recording their street and house number in their rent books—and assigning a number if none existed. The rents were collected mostly on a quarterly basis, with the week's takings given in to the company's office. If a tenant moved, the old name associated with the house was retained until the head office approved the change of name. If a tenant ran away and fell into arrears, the name was passed along to the head office, but the collector remained liable for the arrears until forgiven by the directors. The collectors had the authority to cut off water to tenants fallen into arrears or who used water wastefully. Because of agreements the New River signed with other water companies, they were not to take on customers owing money to other companies. The collectors also oversaw how the water was distributed in their districts. If they judged new pipes were needed, they could order them installed for lengths of under 20 yards (18.3 m) but needed to report their decisions to the board if the modifications ran to more than 12 yards (11.0 m). Anything over 20 yards required the board's approval at the weekly management meeting. They were responsible for signing the pipe crews' weekly reports on work done. In addition, the collectors were also the link with the municipal paving commissioners, who were always sensitive to timely repairs to streets. They signed the company's paving vouchers, reporting work effected in the streets to the commissions.[155] The collectors were paid on commission, taking 5 percent of the rents they collected.[156] Based on total annual rents of around £35,000 in 1770, this translated to £100 to £125 per year per collector. The company's laborers would make around £30 per year during much of the eighteenth century.

Although piped water into the home was the most common way to get this commodity from water companies, it was not the only means. Sometimes, people would pay to supply their stables behind the house, and others living nearby would help themselves to what was available. A similar situation also occurred in courts and small alleys featuring smaller houses. Rather than providing a connection to every house, the water companies at times agreed with all the people in the court or alley to provide a public fountain or valve, in exchange for a fee to be shared between all. As with the stables, this was open to abuse, and many people helped themselves to water from these public valves. In 1715 the New River estimated it lost around one thousand pounds per year from public valves in alleys, stables, and courts. Although some users taking from these semipublic distribution points paid, others did not bother.[157] The Chelsea Waterworks also complained of people taking water from common cocks without paying.[158] In another case, people were discovered to be using carts to draw water from a reservoir. They were threatened with prosecution.[159] Another form of theft involved people selling water to third parties from their own homes.[160] Finally, as just discussed, people were also willing to tap into water company pipes surreptitiously, attaching a pipe to a main passing nearby. In one case in 1803, even a brewhouse was discovered to be taking water without paying.[161] Another way of stealing water was to force out a fireplug. The keys for these plugs were kept by the local parish and the company, but they could be broken, allowing water to flow. This was serious vandalism, which the companies pursued as criminal acts.[162] Another such case involved a lunatic asylum, whose servants were found to be pulling a fireplug to take water from a main.[163]

As central as consumers were for the New River and the other London water companies, they did not take all interested parties willing to pay their fees. The idea that water supply is a necessary service that must be provided regardless of the circumstances was not present in the eighteenth century. The relationship between the companies and their customers continued to be a commercial transaction that could be exited or refused by either party. For the water companies, offering service was predicated on the financial viability of the proposed service. A new customer on a street already supplied with water was always welcome, and the companies sometimes even marketed water to unconnected houses in their supply zones.[164] In areas where no pipes yet existed, the calculus was different. If a cluster of houses on a street applied to a company, then they would be added as customers if the directors judged the potential revenues adequate. Usually, these new customers were asked to pay a portion of the cost of laying the pipes, such as in

Moorfield in 1788. Many people there wanted to improve their supplies, and the New River company agreed to lay a new main, provided they paid half the cost.[165] In another case from 1764, the Chelsea Waterworks got into a long dispute with Elizabeth Chudleigh, Duchess of Kingston, over her refusal to pay ninety pounds for the long pipe extension to her house.[166] Another consequence of the commercial nature of the relationship with customers was that water companies were willing to cut off customers in good standing if, due to remoteness from other customers, they were too expensive to continue to serve. In 1799, for example, the New River directors ordered that a branch be shut off and left for dead, because there was only one tenant on the pipe. What this customer was to do was not stated.[167] The state of affairs did not have to be as extreme as being reduced to a single tenant. The same company in 1804 was willing to plug two pipes because they served only seven out of seventeen houses on one street and six out of eighty-five on another.[168] These ways of operating caused some significant bitterness as an internal New River report related in 1814: the company's unpopularity "was increased by frequently compelling the parties applying for water, to bear a proportion of the expense of laying down the pipes, in some cases refusing to serve, though as a commercial body you were completely justified in doing so."[169]

Manufacturing Pipes

By the late eighteenth century, the New River Company's network of pipes and mains had grown to approximately 400 miles (640 km) in the city (figure 4.6).[170] The company had grown throughout the eighteenth century and continued to lay pipes and take on new customers to 1805. During this time, almost all its pipes were made of wood that rotted out after between twenty and forty years, meaning the company had in effect to rebuild 2.5 to 5 percent of its network every year just to maintain the status quo, without accounting for expansions or reconfigurations. Maintaining this massive network and its continuing expansion therefore entailed a variety of operations to manufacture and lay approximate 10–25 miles (16–40 km) of pipes per year.[171] This required a level of logistics that the company honed over time, including maintaining a supply of wood that eventually stretched to Scotland. In the earliest years of its operations, it had purchased wood locally from farmers or landowners who were willing to part with some of their elm trees.[172] As the pipe network grew, the company developed contacts with timber purchasing agents who worked throughout the country. The company would negotiate

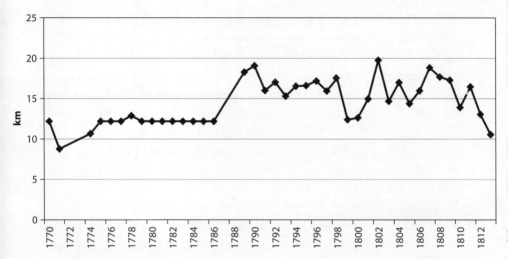

Figure 4.11. Bulk wood purchased from merchants by the New River Company, 1770–1812, measured in kilometers

with them every year for several hundred loads of elm logs to be delivered to its wharf on the Thames. The contracts were often for about two hundred loads per merchant per year, with several merchants receiving a share of the annual total required. The size of each load varied, but the standard was that it they should contain an average of 50 feet (ca. 15.2 m) of unbored logs each. This was an average, and the number of logs per load varied between two and six, depending on their diameters. For most of the eighteenth century, the preferred wood was elm because of its toughness and even grain. The quantity of wood purchased by large yearly contract was steady, usually more than twelve kilometers per year to 1789 and then increasing notably to around eighteen kilometers in 1810. In addition to these bulk purchases of raw logs, the company also bought bored pipes from other merchants and would often purchase trees opportunistically when good deals presented themselves. For years in which the total length of pipe laid is available, these extra purchases represented about 50 percent of all pipes laid.[173] This suggests that the company was buying twenty-five to thirty-five kilometers of wood logs for pipes per year, or 1,600 to 2,300 loads (see figure 4.11).

The price of timber during the eighteenth century was exceptionally constant. Direct data from the company exist from 1770, showing that it paid between 50 and 60 shillings per load, leading to total expenditures on wood of between approximately £4,000 and £7,000 per year.[174] Purchases from the 1680s and 1690s show that it was paying around 40 shillings per load a century before, or roughly a 0.4 percent yearly price inflation over the eighteenth century. Data about English timber prices compiled by economic historians confirm this trend: from around 1660, prices ranged between 40 and 60 shillings per load until the late 1790s, when they climbed, peaking at 150 shillings per load in 1810.[175] From the company's own calculations, it cost a further 2–4 pence per yard to bore the pipes.[176] This meant that the New River and the other water companies faced little pressure from changing timber prices on their model of building large networks of wooden pipes before 1800. Since wood was by far their largest material cost, the basic model was sustainable.

All this changed for the New River after 1800 as prices shot up to over 100 shillings in 1805 and reached 150 shillings in 1810, driven by the disruption of the Napoleonic Wars. The company then became desperate to find wood, buying more of the cheaper fir rather than elm and advertising for timber throughout the country, including in Scotland, where it had not done so previously.[177] The wooden network model effectively crumbled due to its expense, causing dramatic changes to the water technology and the structure of London's water industry. It opened the door to new competitors using iron pipes, which had become available in large quantities at the end of eighteenth century (fig. 4.12). The introduction of new techniques for iron production, notably coke smelting and puddling, vastly expanded the quantities produced and the possible uses for iron.[178] Furthermore, in 1784 the Chelsea Waterworks' engineer Thomas Simpson invented a new way of joining iron pipes. Previously, they had been attached end-to-end by flanges. The problem with this arrangement was that, with no room for expansion, changes in temperature caused the pipes to break. Simpson devised a joint whereby one pipe was inserted into the next one via a socket. The joint was sealed with hemp or flax and soldered with lead.[179] Iron could then be used in greater quantities, and new competitors dislodged the New River from its leading position when the wood price spike came. Wood was entirely displaced by iron in the London water industry within a few years of 1810, as is described in chapter 8.

This shift to iron pipe in the early nineteenth century did not mean that iron was not used before. Although the transition to iron pipes came after 1800, the material was used to make other components earlier, even from very beginning. Pipes were bound together under the streets using iron hoop as early as 1613.[180] With the increasing availability of iron in the 1780s, however, the company experimented with iron valves, whereas they had been

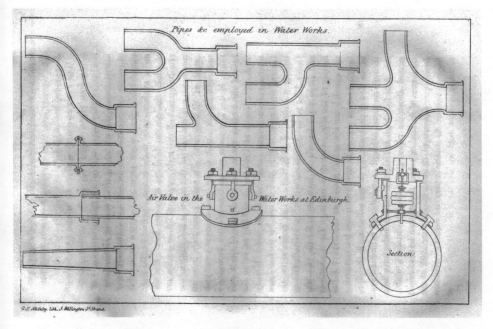

Figure 4.12. Iron pipe joints, turns, and splits. William Matthews, *Hydraulia* (1835), p. 143

made of wood or brass from the early days.[181] Within a few months, the directors decided that the iron valves were so successful that they no longer needed half the laborers they employed to bore wood valves and laid them off.[182] A few months later, they resolved to use only iron valves, replacing even brass ones.[183] Another iron component tried at the same time was a cylinder inserted to join wooden pipes together.[184] The development of the British iron industry in the late nineteenth century therefore had important consequences for London's water industry.

Once the wood logs were delivered to the company's pipe yard, they had to be turned into pipes (fig. 4.13). Most of the pipes were 3 to 5 inches (7.6–12.7 cm) in diameter, while the mains were of 6 to 8 inches (15.2–20.3 cm). The company did not buy large trees as their trunks were too wide and their branches too gnarled to serve for pipes. As the wood arrived, they were bored out by several crews of pipe borers who were paid a piece rate on a rising scale, from three pence for a three-inch pipe, up to five pence for a five-inch pipe. Pipes of larger bore were prepared by horse engines.[185] The pipe had to be bored carefully so that the exterior walls were not too thin, leaving them liable to rupture.[186] The company employed around twenty pipe bor-

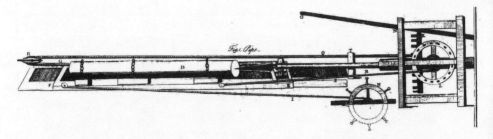

Figure 4.13. An eighteenth-century pipe-boring machine. The log was mounted on a sliding platform and was driven into the auger by a system of pulleys. *The Complete Dictionary of Arts and Sciences* (1765), vol. 2, pl. 150

ers working full time at they yard, and they were all overseen by a foreman who lived on site.[187] The sections of pipes were narrowed at one end, while the opposite had a larger hole so that they could fit together snugly end to end. The company used metal hoops to bind the end of a pipe to prevent its cracking when another was being inserted into it.[188] Three-way splits, where a pipe branched into two others, were achieved by means of junction boxes with three holes in it, one for the inflow, two for the outflow pipes. These three-hole pieces as they were known were made of brass or even wood.[189] With 25–35 kilometers of pipes being laid every year, the pipe yard produced about 80–120 meters of pipe per day, assuming they worked six days a week the entire year. Given this volume, the company sometimes explored ways of boring more rapidly, such as in 1801 when the directors looked into a kind of boring machine devised by William Murdoch.[190] Murdoch was an engineer working for Boulton & Watt who was central to the early history of gas lighting. Another simpler attempt at raising efficiency occurred when the company paved the circular paths the horses trod as they turned the machines.[191]

Maintenance

Producing pipes was one step, but getting them into the ground was an even larger job that was part of the ongoing maintenance of the company's infrastructure. Maintaining and building the company's vast infrastructure of mains, pipes, valves, junctions, and other components constituted the single-most demanding class of activity the company engaged in, from the point of view of number of employees dedicated to the task. Much of the maintenance consisted in finding and fixing faults in the pipes, mostly leaks. Among these the easiest to find, but also the most damaging, were main ruptures.

Due to the volume of water they carried, burst mains would often wash away part of the road, such as happened on Goswell Street in 1770. The street was a turnpike under the care of a trust, and after the incident the trustees sued the New River for causing a nuisance and claimed damages. The court rendered a verdict against the company.[192] Bursting mains also typically damaged cellars, which became flooded with water, but there could be other victims too.[193] In 1795 the company paid two guineas in compensation for nine pigs drowned in a cellar when a main burst in Wentworth Street.[194] At other times, water rushing in the street could damage walls of buildings, such as the Southampton Buildings in Chancery Lane. In these cases, the company agreed to pay all expenses.[195] To limit interference with cellars, the company would move pipes from the footways along the sides of the road toward the carriageways in their center, increasing the likelihood that leaking water would stay in the road.[196]

Catastrophic main ruptures may have been damaging, sometimes dramatically so, but given the constant inexorable degeneration of the wood pipes, slower leaks were a much more frequent problem. Reports of complaints from aggrieved householders were common throughout the company's minutes.[197] For example, in 1782 two people in London Street complained that a wooden pipe above their wine cellar had leaked water, causing extensive damage. The pipe was replaced with a lead one.[198] In a second instance, an old service pipe "in very bad condition" was drowning cellars along Finsbury Pavement for some time before it was replaced.[199] In this case, however, the company declined to pay for all water damage. In 1799 Jonathan Hoare in Stoke Newington requested reimbursement for a bricklayer's bill for repairs effectuated after water entered his cellar. The bill was refused because it happened when the pipes were being laid in the street at the inhabitants' request, and they had been informed that "probably they would be incommoded in that way."[200] In other times, long sections of pipe became so degraded that finding a single leak would not noticeably improve the situation. This happened in 1796 when many people on Pitfield Street complained of defective service due to leaking pipes. After investigating the situation, the directors decided that the entire main should be replaced.[201]

Finding leaks could be very difficult. Complaints about poor water supply could have any number of causes, but if water was seen running in nearby sewers, the company could be fairly sure that leaking was the cause. Tracking down where exactly the hole in the pipes was could be very distasteful, and even quite dangerous for the laborers. They would enter sewers through grates and move along them underground, seeking the spot where the water was entering. Entering foul sewers in this way was so repugnant that the laborers were for some time paid a bonus of one shilling sixpence every time they did so. Their usual daily salary was two shillings.[202] The directors, however, decided in 1786 that this was excessive, and the bonus was paid only if the worker had to go more than one hundred yards in the sewer.[203] The perils of this work were all too evident from some of its tragic results. In 1794 a laborer died in a sewer searching for a leak.[204] Others fell ill from going into the sewers or were injured.[205] Moreover, leaks did not always go into sewers where they could be hunted down. In some cases, although the company was aware of a loss of water somewhere in a local section of the network, locating the actual source could be very difficult.[206] In 1786 a loss of water from the service pipe in Blackmoore Street was noted, but the hole could not be found. The company shut off the service, leaving the tenants without water. After a few months, the hole was finally found, and the service pipe's source was shifted from Drury Lane to Stanhope Street.[207] The directors' expectation was that the foremen responsible for a district find and repair all leaks. If they did not, they would be reprimanded, as happened in 1788 to a foreman for allowing fault on the pipes to persist. He was ordered to hire two more laborers, one at his expense, to mend them all.[208]

Supply problems were not caused only by leaks. All sorts of other factors could be to blame. One was that dirt could enter the pipes, which then became clogged with mud.[209] If too much mud accumulated in a pipe, it sometimes had to be replaced entirely with a new one.[210] The company took preventive measure against the accumulation of detritus in the pipes by installing a plug at the end of every pipe and main in the network. Over the course of a month, the foreman would open all the end plugs in his district and let the water run to flush the pipes. Air in the pipes could be even more of a problem because it was much more likely to enter the network. Air pockets caught in corners could cause a significant reduction in the flow. This was so much of a problem that the pipes were described as "wind-bound" if they locked up this way.[211] In 1763 the article on "pipe" in the *New and Complete Dictionary of Arts and Sciences* claimed that in practice, air in pipes was the chief impediment to water flow. The solution suggested there was to drive a nail into the pipe to allow the trapped air to escape.[212] An earlier solution suggested in the 1720s came from John Desaguliers, who proposed linking a small secondary pipe to water mains to allow air to escape from them through a valve.[213] The New River, however, devised other means to deal with wind-bound pipes. It installed air plugs at regular intervals at the high points of the mains. Every week, each one of the air plugs was opened to allow any air that had found its way to this local peak in the network to escape.[214]

Some problems with flow, such as kinks and irregularities, were caused by factors not remedied by opening valves. Because the pipes lay under city streets, some with heavy foot and carriage traffic, they were subject to regular pressure from above, causing them to shift and leak. This pressure was greater the closer the pipes were to the surface, and many times the company would re-lay the pipes in the same street at a greater depth to avoid damage.[215] The pipes could also become very irregular in the course they followed. Although soil pressure could result in some shifting, more frequently it was simply the way they had been laid, with corrections and additions to short sections leaving a haphazard path. For instances, collectors in 1786 reported that because the Clare Market main was very irregular, especially in Kingsgate Street, many sections had to be taken up and corrected.[216] In another case from 1804, complaints of poor service were received from customers, and the street inspector examined the pipes in Brick Lane in Spitalfields. He determined that while the mains were in good shape, the service pipes were not. They needed to be re-laid with new wood along a straight path and well below the surface to avoid damage from the pavement soon to be laid there.[217]

At times the problems were inside houses, such as having the cistern too high, which could prevent the water from flowing in.[218] At other times it was simply that the customers were prickly. In 1781 a tenant in Tichfield Street was "greatly distressed" about her many years of poor supply, as well as the "rudeness and impertinence" of the turncock. Investigations showed that the problem lay with the small size of her ferril, and the placement of her reservoir. Both of these were under her control, but she refused to pay to change them. Moreover, the turncock was reported to be a "very civil obliging man."[219]

Freezing was a seasonal problem. In times of deep frost, if the pipes froze, water could not be supplied in some zones. The lead ferrils would freeze first because they were the smallest pipes and were often exposed to the ambient air where they entered buildings. To warm them during cold periods, some people would pile dung on the road above the pipes, as well as on the ferrils leading into the house. The efficacy of this measure was debated.[220] If the cold was deep and prolonged, the service pipes and even the mains could begin freezing, at first where the ferrils were attached. The supply would then fail altogether in places, and the company would attach standpipes to the mains to distribute water. Standpipes were sections of pipe a few feet long that could be screwed into the main where fireplugs were located. They had valves on them, allowing people to take water from the mains, which, being the largest pipes, froze but rarely. The standpipes were used until the freeze passed. There were, however, some occasional problems with the standpipe solution. In 1788, the standpipes were left open leaving water to run onto the streets with the result that sheets of ice formed to the dismay of those trying to walk there.[221] At other times, if the standpipes were slow in coming, people took the matter into their own hands by cutting into the mains.[222] Some severe frosts could damage valves and pipes enough to necessitate replacement.[223]

The deepest colds presented an entirely different class of challenges. In 1795 the aqueduct's freezing caused water to stop flowing to the entire network. A number of mains then froze in the area close to the New River Head where they were above ground. Normally, the constant flow in that area prevented this from happening. The freezing happened in stretches of pipe too long to break up. These turned out to be damaged and had to be replaced. Laborers were sent out into the city to set up standpipes to distribute water where the mains were functioning, but many were split and froze onto the street. The local parishes were forced to break up the ice to remove it and claimed compensation for their trouble from the company. The company rejected the claim, stating it could not be avoided.[224]

All this work of investigating problems with supply, finding leaks, digging trenches, laying pipes, and paving the streets necessitated a large operation, bigger than any other part of the company's activities. It employed twelve teams of laborers, each headed by a foremen or "paviour," with four or five men per team.[225] These teams worked twelve hours per day six days a week, with occasional overtime. In addition, there were two carts each with a driver and two horses constantly hauling paving stones, pipes, and debris throughout the city, keeping the teams supplied.[226] Much of the work they did was in reaction to the many problems that were constantly springing up. When a fault of some sort was first observed or reported to a collector, he took note of it in a fault book he kept. The company's policy was to fix faults within forty-eight hours of being reported.[227] After hearing of the fault, the collector would then work with the foremen of the district to identify and mend the fault.[228] The fault books would be sent to the head office once per month for examination.[229]

The foremen, working with the collectors, were the key figures in the maintenance operation. Like the turncocks, they had a long list of clearly defined and enumerated responsibilities. Their role was to investigate all problems and make sure all tenants were "equally and sufficiently supplied." A foreman was expected to be "fully acquainted with all mains, service pipes, cocks, and everything within his district" and should also "keep in his pocket a copy of all the turnings, made by the turncock . . . and if he sees any neglect or alterations in the regular course and succession thereof," he was to report it. He had to survey all the mains and riders every week up to the New River Head. He was to remove the end plugs of all pipes at least once per quarter "to scour and clean them." He also had to open the air plugs at least once per week to let out the air that could wind-bind the pipes. If a deficiency was reported, the foreman would do tests in various parts of his district by inserting a "cane" or small pipe into the mains at various points to see to what level the water would rise. If a problem was identified, he and his workers would enter sewers, vaults, and cellars seeking possible signs of a leak. When work was needed, the collectors could approve minor changes, but larger ones required the board's approval.[230] The foreman was responsible for setting up standpipes in fireplugs during freezes and ensuring that they were never left unattended. The most common work foremen did was laying pipe. They and their crews would dig trenches and cut the pipes from the yard down to the necessary size. The old pipes and scraps were sent back to the yard and not to be "given away for beer." When installing ferrils, none were to be placed on the main, or on service pipes above its valve linking it to main. The foremen were supposed to inspect all buildings where they suspected too much water was being used. They were also to examine the uses the tenants made of the water. Like the collectors, they also had to live in the district in which they worked. They were paid four shillings per day.[231]

In 1795, the company attempted a greater degree of centralization of decision making and information collection by creating a new post of street inspector that removed some of the responsibility from the collectors and foremen. The inspector had oversight for the entire city, not a single district. To make this possible, he was mounted on a horse and was supposed to ride throughout London, looking into complaints, directing the foremen's work, and reporting back on the state of the company's work throughout the city to the directors. The post was maintained for many years.[232]

With all the different things that the pipe crews did throughout the city, the most common was that of was digging trenches and laying pipes. Given the large number of crews the company had and the length of pipe laid per year, this meant that open trenches were a common sight in London in the eighteenth century. When the work could not be completed during a single day, the trenches were left open overnight, creating a potential hazard in dimly lit streets. The crews were supposed to leave lanterns to mark the open trenches from 3 PM in the winter, but there were frequent cases of people, horses, and carriages falling into them, sometime even killing horses.[233] The old pipes removed from the streets were all sent back to the yard, where inventory was tracked daily. Some of these pipes could be reused, while the rest were sold.[234] The pipes had to be laid in the mains tightly. If the ends were not snug, they would have to be re-laid.[235] Given how much time was spent digging trenches and laying pipes, the company occasionally experimented with means to speed the process, as it did with pipe boring. In 1752 the company's engineer Henry Mills devised a new method for laying pipe that the company adopted. Although the details are not clear, it made for better connections between the pipes, extending their life, and was useful for "preventing so frequent changes in the streets." It required new instruments for the "gaging" (or measuring) of the pipe joints.[236]

Legal Dimension

There were also legal and political aspects to running the New River's infrastructure network. One of the most important was protecting the company's property. Its control over the river as well as its reservoirs and pipes was often caught within a context of overlapping property claims. Over the years, the law, particularly through the courts—but also via royal proclamations— was willing to protect the New River's property against competing claims and encroachments. Without this protection, the desire of other parties to draw water from the aqueduct itself or, more importantly, to deny the company access to their own land by removing the company's pipes or reservoir infrastructure may have crippled the company. The law's support for the company's ability to function on property belonging to others began with the original acts that created rights to water and then with compulsory access to land in order to build the aqueduct. Throughout its entire history, the company was secure in knowing that it stood on a solid legal basis for running the aqueduct through land not of its own, however much some landowners may resent it. The only ambiguity was that these first water rights had been granted to Hugh Myddelton personally rather than the company, which had not been incorporated yet. The company had unsuccessfully tried to remedy this issue by getting its own acts of Parliament, but in practice this detail did not pose it any problems over its history.[237]

More serious challenges came from landowners who wanted to remove the company's pipes or reservoirs from their land because no land access rights for pipes as opposed to the aqueduct had ever been explicitly granted to Myddelton or the company. For its part, the City of London never seems to have challenged the New River's access to the streets, nor did the various paving commissions. Private landowners, however, were a different matter. On a number of occasions, landowners who previously agreed to allow the company access to their land changed their minds or demanded much higher payments. In some cases, such as the Duke of Northampton's fields at the New River Head, rerouting the pipes or moving the reservoirs would have been impossible. The first time a dispute of this sort went before the courts was in *New River v. Henley & Baynes* in 1694. The company had eight mains crossing Cold Bath Field toward Great Ormond Street and on to Westminster. The lord chancellor at the time, John Somers, ruled that, because the water supply was a public good, the landowner could not arbitrarily remove the pipes or charge any arbitrarily high fee. To resolve the dispute, he stipulated that an independent committee should establish a reasonable price that the landowner could charge the New River for access to his land, in effect following the model used for compulsory land access when the aqueduct was built. The committee decided that eight pounds per main per year was reasonable, and although the landowners objected, Somers upheld the decision.[238]

Another case, *New River Company v. Graves*, came in 1702, this time over three mains running through fields close to the New River Head. The company had signed an agreement with the previous landowner, but not long before the lease expired, Graves purchased the land and tore up the pipes. In his arguments before the court, Graves claimed that the company had been granted compulsory access to land only for its aqueduct, not for pipes. The lord keeper Nathan Wright, adopting Somers's view that the water supply represented a public good that should be protected, rejected Graves's argument. He wrote that "the act is to be taken in a liberal sense that the Town in general might be served with water."[239] Furthermore, "the bringing of water to the reservoirs would be of little use to the town unless the company had liberty to distribute the same by pipes." A commission was once again appointed and ruled that the company should pay five pounds per main per year.[240] This sort of precedent established around 1700 effectively gave the New River Company, and by extension the other water companies, legal cover to function as large infrastructure networks that relied on other people's property. This was, in effect, a legal model for a special kind of business activity. As described in chapter 1, compulsory access to or expropriation of real estate was not new, but as the tremendous public service the water supply companies provided became clearer with time, the compulsory access rights were explicitly recognized as meriting special treatment in the law that bypassed ordinary negotiations between private parties. This applied even in cases that were not explicitly outlined in the relevant charters and acts.

Conclusion

The New River Company's achievement in supplying ever greater numbers of customers after the Great Fire and then over the course of the eighteenth century was much more than one of laying more pipes and acquiring more water. Indeed, the sources of water it had access to in 1800 were the same as 1650: the springs of Chadwell and Amwell and the River Lea. The company had faced a crisis of supply in the late seventeenth century when, seeking help in resolving the problem, it turned to Christopher Wren and John Lowthorp. Both of these men indicated that a systemic solution was needed. By measuring the aqueduct's flow rate, Lowthorp proved that the problem lay not with the quantity of water brought to London but with what happened to it once there. Wren, using organic metaphors, argued that the company's works should be seen as body that functioned as a whole. The entire system needed to work as one, and patchwork solutions would not do. They both indicated that the company had an inadequate knowledge of its own works and needed to reform its approach to management. Lowthorp specifically recommended creating the post of surveyor with specific responsibilities for the whole works. Although it is unclear to what extent the specific suggestions Wren and Lowthorp made were at the root of the system that emerged, some of their observations were clearly echoed in what came afterward, such as the post of surveyor. Building on many elements that were already in place, the company developed and refined an approach to stabilizing and maintaining its network that proved effective. Most of these elements together aimed at increasing the efficiency of the distribution of water, and getting it to where it could be sold for the greatest profit. The range of measures ran from getting and storing enough water at the reservoirs, to distributing it daily via scheduled turning of valves, to managing load at various levels, to controlling users and removing obstructions. No one appeared to have formulated a network philosophy, at least in any archive or publication that still exists. Nevertheless, the New River had created a carefully implemented and centralized system to stabilize its network. This system relied on a few hundred people working daily to produce pipes, turn valves, collect fees, and dig trenches. The directors had implemented means to control all these people, via maps, inventory lists, books of turnings, and rent rolls. Even the places where its turncocks and foremen lived and traveled were controlled.

The practices that the company put in places were supported, on the one hand, by a favorable legal environment and, on the other, by the broader economic conditions. Legal support was based on the opinion that because it was providing a great public good, the company should be granted access to land on reasonable terms, even if that right had not been explicitly granted by charter or act. Court cases around 1700 established this, and although some people threatened the company over the years with extortionate rents to access land, the directors could be secure in the knowledge that, if it ever came before the courts, they would rule in their favor. The economic environment was one of great stability over the course of the eighteenth century. The price of wood, the New River Company's great material cost, was unwavering and posed no threat to its business and technological model. Only after 1800 did this begin to change, and what occurred then demonstrated just how dependent the model was on relatively cheap and available wood. The huge run-up in the price of wood after 1800 shattered the New River's model of water supply, so that London's water industry was dramatically reshaped, dethroning the New River from its perch, and forever banishing wood as the prime material for pipes. The technological model shifted decisively. The price of wood was not, however, the only force. The expansion of the iron industry with new means of production pointed in a new direction, and the water industry was carried along by the broader economic changes in the British economy that was the industrial revolution.

Natural and experimental philosophy, or at least people interested in it, had some impact on the water industry. Wren and Lowthorp, both members of the Royal Society, had a role in reshaping the systematic design of the New River's network. The loss of archival material makes it is difficult to determine definitively a direct influence of their recommendations on the subsequent history of the New River, even if some of their recommendations were implemented with time. It is even less clear to what extent their interest in natural and experimental philosophy contributed to their recommendations. Nevertheless, the New River directors looked for external advice from people highly regarded for their intellectual and design skills at a time of crisis. It was the Baconian dream of the founders of the Royal Society that their natural philosophy should be useful for ameliorating the human condition. Wren's ambition for the reconstruction of London on such a rational plan after the Great Fire had come to naught, but he laid some seeds for a systematic approach in the New River Company, as had Lowthorp. Later in the eighteenth century, the hydrodynamics work of the French natural philosophers Dubuat and Bossut in exploring the effects of friction on pipe design also may have had an effect as the New River made design decisions about pipes that seemed to have flowed from their work, although any intellectual debt was not made explicit in archival sources. Furthermore, to the extent that steam engines had some links to natural philosophy, they became ever more important in the London water industry, pumping over 50 percent of supply sometime around 1810. All these links are, however, tenuous or indirect. Most of the technical innovation in the water industry—probably the vast majority of them—were connected not to natural philosophy but rather to the contributions of mostly anonymous workers who developed usable ballcocks, iron spigot joints, and the like in the course of their daily jobs.

CHAPTER FIVE

The London Bridge Waterworks and Other Companies in the Eighteenth Century

From the year 1700 the time when these Works became a publick Stock, such Additions, Alteration, and Amendments have Occasionaly and Experimentaly been made from time to time in the Wheels and Engines as well as in their Apparatus that it may now very Justly be esteem'd a Compleat peice of Machinary, Constant in its performance and as Extensively usefull as any in the whole World.

Samuel Hearne, *Report to the directors of the London Bridge Waterworks*, 1745

Although the New River Company dominated the London water industry until the early nineteenth century, its difficulties in the late seventeenth and early eighteenth centuries demonstrated that it was not in an unassailable position. After it regained its footing, its profitability recovered and then far outstripped that of the other companies, but they nevertheless survived. A combination of technological, business, and geographic factors allowed them to do so. The most important of these factors was that the New River was limited in the amount of water it could supply. Although it regularly sought to increase the volume of water at the New River Head, it was fundamentally constrained by what it could draw from its springs and the River Lea, in addition to the challenges in distribution outlined in chapter 4. Furthermore, since the New River made only limited use of pumping technology, its supplies could not reach more elevated areas along the Thames into Westminster. It did not introduce more pumping than it did because, lacking extra water supplies and with ample potential market already available, it simply had no reason to reach these areas. A second factor allowing other companies to exist and grow was that the demand for water was sufficient enough that these companies could hold their own in smaller zones of supply, even when in direct competition with the New River. The London Bridge Waterworks was the preeminent example of this. Although during the seventeenth century the two companies overlapped somewhat in supply areas, by the mid-eighteenth century the New River was selling across the entirety of the LBWW's supply zone. Despite this, the LBWW kept its prices low enough that it supplied more houses in its area than the New River did. It managed to do this in part because it invested heavily in its infrastructure during the eighteenth century, increasing the quantity of water it could supply. It rebuilt its water pumps several times and improved how it distributed water through its network.

This chapter explores the history of the other London water companies in the eighteenth century, especially the LBWW as the largest one after the New River. Its history was heavily influenced by where it was situated, as was the case with the New River. The ability to use gravity to supply water was one of the New River's advantages over its major competitors. With the exception of the Hampstead Waterworks, all the other significant companies had to pump water from the Thames using waterwheels or, later, steam engines. Pumping technology, therefore, was central to their ability to operate, and it was this technology that had originally catalyzed the birth of the new water industry in the form of the LBWW. The LBWW's defining feature was its position on London Bridge where its pumps were located. The size of the bridge's arches imposed a limit to how much water could be pumped, and the current there determined the rhythm of its pumps. The only way for the LBWW to increase its capacity was to improve the waterwheels or to occupy more arches on the bridge. The only other option for the LBWW was to expand off the bridge, something that it managed to do to a very limited extent in the eighteenth century. The company's reorganization after 1700 proved to be a turning point. The LBWW's new owners, with more capital at their disposal than the Morris family, were willing to invest in the company. They repeatedly rebuilt the waterwheels over the course of the century, seeking to improve them. In addition, they occupied more arches. This allowed the LBWW to keep its prices low enough to fend off the New River.

By contrast, the third largest company, the Chelsea Waterworks, did not need to face the New River in the same way. Compared to most of the other companies, it was a newcomer, having been founded in the 1720s. Its primary pumps were tidal, situated in Chelsea upstream from Westminster, although it later introduced steam pumping as well. Much of its supply areas in Westminster and Chelsea were not served by the New River, giving it scope to establish itself. It did so with some difficulty, repeatedly seeking capital from shareholders and lenders. Had it faced direct competition from the New

River, it would not have grown as fast. After its initial growth spurt, it grew only very slowly and was barely profitable, if at all, up to 1754.

Business factors allowed both the Chelsea Waterworks and the LBWW to survive and expand their infrastructure. In contrast to the New River, which hardly ever had to issue debt or make capital calls on its shareholders after the first decades of its existence, the Chelsea Waterworks and the LBWW relied a great deal more on external capital. Besides the capital contributions shareholders made in the early eighteenth century, both of these companies borrowed money to improve and expand their works. The joint-stock form made this sort of large-scale fundraising from shareholders possible. Being a corporation seemed to be less critical however, as the LBWW was not hobbled in this regard by its unincorporated status.

The Nature of Competition: Dominance of the New River and the LBWW

Although the New River was far larger than the other London water companies combined, some of these secondary players nevertheless attained a fair size, especially when compared with water companies in other cities. Despite the loss of much of the archival material from secondary London companies, enough remains to gain an idea of their relative sizes. The New River's growth shows that it already had more than ten thousand customers by 1670 (see fig. 6.1). No other company reached anywhere near this figure before the late eighteenth century. The LBWW was the second largest, and its growth was especially made possible after the chaos of its reorganization into a joint-stock company around 1700 had passed. Around 1710, an investor estimated its revenues at forty-eight hundred pounds, meaning it served approximately forty-eight hundred customers because its agreement with the City limited its rental charge to one pound per annum.[1] After the waterwheels were subsequently rebuilt and new bridge arches leased, the company reached around eight thousand houses in 1745.[2] It thereafter grew more slowly, in 1780 possessing ten thousand customers, a figure that was not significantly higher even around 1820, just before the bridge was destroyed.[3]

Between them, the LBWW and the New River dominated the water market. According to some statistics, such as volume of water supplied and houses served, they took roughly 80–90 percent of the market for most of the eighteenth century.[4] The total dividends paid over the years also show how much the New River dwarfed the other companies (fig. 5.1). If the New River was larger than all the rest together, the LBWW was in turn about the same

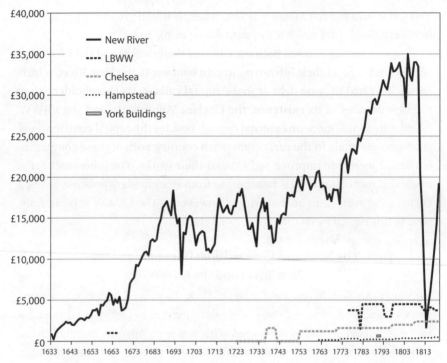

Figure 5.1. Comparison of total dividends paid by the water companies

size as the remaining companies combined until 1750, when the growth of the Shadwell Waterworks and the Chelsea Waterworks combined started to catch it. The Shadwell even passed it in terms of number of houses served sometime in the 1780s. Among the smaller companies, the third largest company was the Chelsea when measured by volume of water served and rental income, but not by number of customers. The Shadwell had more customers than the Chelsea by the end of eighteenth century (see fig. 5.2, as well as fig. 6.2), but because it served the poor East End as opposed the Chelsea's wealthy West End, the volume of water it served was about 40 percent of the Chelsea, and its gross rental income was 70 percent.[5]

Like the Chelsea, the York Buildings Company initially acquired tenants quickly, with twenty-seven hundred in 1691.[6] It did not, however, enjoy the same constant growth that the other company displayed over the long run and had a number of false starts (for its attempts with steam engines, see chapters 3 and 4). After 1800, it had about two thousand customers.[7] Its water assets were taken over by the New River in 1818.

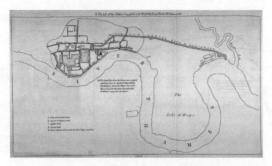

Figure 5.2. Map of Shadwell Waterworks' mains in 1795. *Report from the Committee Appointed to Enquire into the Best Mode of Providing Sufficient Accommodation for the Increased Trade and Shipping of the Port of London* (1796), appendix E

TABLE 5.1
Number of Mains in 1739

Company	Mains
New River	58
London Bridge Waterworks	8
Chelsea	5
Hampstead	2
York Buildings	2
Marchant's	3
Hyde Park	3
Shadwell	2
Marylebone	1
Rotherhithe	2
Bank End	1
St. Saviour	1

Two further data sets provide a good idea of the relative sizes of the companies. In 1739 William Maitland listed how many principal mains each of the companies had (table 5.1). This showed that the New River had fifty-eight mains (figs. 5.3, 5.4), while all the rest combined had thirty. The number of mains gives some idea of the size of the company's network, but since water was not supplied continually, the correspondence was approximate. The smaller companies in particular likely used less of the capacity of their principal mains than the larger ones. Further information on the size of each of the water companies dates to the 1770s, when Robert Mylne, the New River's chief surveyor, reported the quantities of water supplied by the city's largest companies. The figures for the companies besides the New River are clearly rounded estimates, but they roughly follow the overall picture presented by Maitland's figure from thirty years before, with the exception

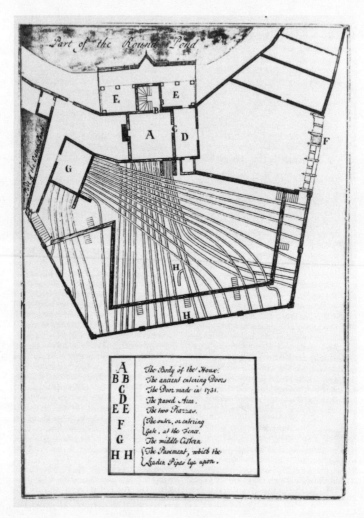

Figure 5.3. Map of the New River Company's mains at the New River Head in 1731

that the Chelsea Waterworks and Marchant's Waterworks supplied less than what the number of mains reported earlier suggests (table 5.2).[8] In the case of Marchant's Waterworks, by the 1770s it was close to exiting the market altogether and being shut down by the City for only grinding grain, so it had evidently declined since Maitland's 1740 report. The reason for this deterioration was that its reservoir had been purchased by property developers and built over. A map from 1746 showed that the company had a reservoir at the north end of Rathbone Place, on the north side of Oxford Street from Soho Square.[9] These areas were built up in subsequent years, and the reservoir

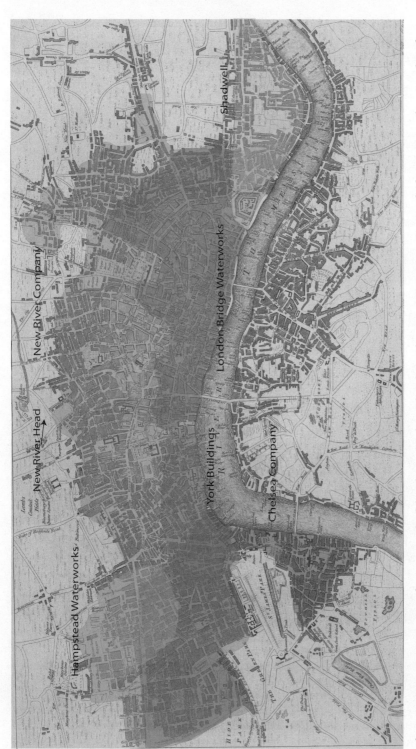

Figure 5.4. Water company supply zones north of the Thames in the mid-eighteenth century. Reconstruction overlaid on *A new and correct plan of London and Westminster and Southwark with new buildings to the year 1770*

TABLE 5.2
Volume of Water Supplied Daily in 1767

Company	Volume (tuns per day)	Volume (ML per day)
New River	57,897	55.29
London Bridge Waterworks	8,500	8.12
Chelsea	1,740	1.66
Hampstead	1,200	1.15
York Buildings	840	0.80
Marchant's	205	0.20

did not appear on a map from 1767.[10] The York Buildings Company suffered a similar fate. It built a reservoir north of Cavendish Square in the 1720s when the Duke of Chandos, John Brydges, was developing housing in the areas surrounding the square. Being a shareholder in the York Buildings Company, he made a deal with the company to supply the area, and it built a reservoir there, known as the Marylebone Basin.[11] It was, however, too difficult to pump water from the Thames to that elevation, and it gave up supplying the area in the 1730s, selling the reservoir to the Chelsea Waterworks (see chapter 4).[12] Its Cavendish Square reservoir too was gone by 1767.[13] These two companies, therefore, had decided that the land on which they had reservoirs was more valuable if sold off for development than as an asset used for selling water.

The water industry developed much more slowly south of the Thames, in part because of the smaller population, but also because it was lower on average and had more wells. A small company, the St. Saviour Waterworks, was founded by 1720, supplying water by tidal mill. This seems to have become the Bank End or Bankside Company, associated with Thrale's brewery. Finally, it was taken over in 1771 by the (Old) Borough Waterworks, which had been founded around that time.[14] The only sizable company on the south side was the Lambeth Waterworks, incorporated by act of Parliament in 1785.[15] Initially, it was small, with thirty-two shareholders each holding a one-hundred-pound share.[16] It grew fairly rapidly, acquiring 3,000 customers by 1800, and 11,500 by 1820.[17]

The New LBWW to 1750

For the LBWW, the eighteenth century was a mixed affair. It had a good beginning because its reorganization into a joint-stock company gave the company access to new capital that allowed it to undertake major projects to renew and expand its infrastructure, especially the waterwheels, which were the foundations of its operations. Although this secured its place for many decades to the 1770s, over time it increasingly bumped into the limits of its situation on the bridge. By the 1780s, further attempts to expand were hindered by the lack of available pumping capacity, and growth slowed and even halted. Although the shareholders were willing to invest more, this sufficed only to maintain a certain level of operations until its definitive end came in the nineteenth century.

For many decades in the seventeenth century, the LBWW had been quite profitable: the company was generating a profit of about one thousand pounds per year for the Morris clan by the 1660s.[18] This was sufficiently high for the family to rebuild after two fires, the first of which occurred in 1633 and partly damaged its works above the bridge. The second was the 1666 fire, which almost completely destroyed it.[19] It was slower in recovering from the second disaster because the lawsuit between various members of the Morris family delayed reconstruction. It did not return to the same profitability that it had enjoyed before the fire, in part because the New River had expanded its reach. This prompted the family after 1700 to sell out to Richard Soame and his partners, who converted LBBW into an unincorporated joint-stock company.[20] The willingness of these new investors to invest capital helped the LBWW to consolidate its position after many new companies had formed. It was still the second largest water company in London, being approximately a third the size of the New River at this time, an estimation based on the number of employees it needed to maintain its works and to collect fees (three turncocks versus eleven for the New River).[21] Some of the more recently formed companies, possibly the York Buildings Company to its west and certainly the Shadwell Waterworks to its east, threatened the LBWW's areas of supply. The small Broken Wharf Waterworks to its east were the closest, overlapping with its supply zone, but it was too small to pose a serious threat. Nevertheless, the acquisitive new owners decided in 1703 to purchase the company from its owner, Benjamin Ayloff, for an annuity of unknown duration, paying three hundred pounds per year. The LBWW took over operation of the pumps there, which was driven by two horses.[22] The New River, however, clearly remained the major threat to LBWW. The possibility of direct competition was also greater at this time because, as described in chapter 4, the New River was consolidating and expanding its own capacity by building the upper pond supplied by a windmill, as well as beginning a reorganization of its pipe network. In 1711 the New River directors specifically decided to target the City to regain customers from the LBWW that it had lost in the preceding years. A recommendation was made to the New River directors that by increasing the frequency of its supply from 4 AM to 10 PM daily, Sundays excepted, it could repulse the LBWW, which served water only every second day and had difficulty in dry times.[23]

The expansion of the LBWW began with the lease of the fourth arch of the bridge in 1701. Once it was in the company's possession, one of the LBWW's new owners' first acts was to hire George Sorocold, the same prolific engineer who had built Marchant's overshot waterwheel and would design the New River's upper pond, to build a powerful new pumping mechanism under the newly leased arch in 1702.[24] Sorocold placed two waterwheels under the fourth arch, with at least one of them featuring his newly patented mechanism that allowed the waterwheel to rise and fall with the tides, with the aim of keeping the water flow steady.[25] He also rebuilt the mechanism under the first arch. Soon, the LBWW was using its expanded supply capacity against the New River in the City, taking tenants from it and causing its revenues to decline. The New River's directors worried about this growing and newly aggressive LBWW, but they thought they could still win back tenants because they could supply water more frequently than the alternating days of the LBWW.[26]

The LBWW's efforts to remain competitive and grow continued over the succeeding decades and included rebuilding many of its waterwheels and increasing the flow rate through the arches in which its wheels were located. In addition, it tried to add even more arches to its collection. In 1731 the owners asked the City to lease the third arch of the bridge, the one between the first and second, and the fourth, where the LBWW already had waterwheels, but this was rejected.[27] The LBWW also tried to fend off competition. It managed to reach an entente with the New River Company in 1738 when the two agreed not to take customers from one another by offering water with no upfront connection fee, nor would they take customers by "false pretences or clandestine means." Only if a customer had no arrears with their current supplier would they allow a transfer.[28] While this was not a collusive agreement that set a floor on prices, nevertheless it removed some of the competition between the two as connection fees could be quite high, even as much as a year's fees for water. The most likely reason why the New River did not press the LBWW more than it did, however, was simply its inability to supply those areas of the City around the bridge with water. In Richard Castle's 1735 report for the Dublin City Council on its water supply, he observed that "probably, had this [New] River been capable of supplying the whole City, the machine at London Bridge wou'd have met with the same fate as that of the York-Buildings, which is now laid aside . . . Nor would the Chelsea Water-Work (tho' the best contrived machine of its kind) have been thought worth pursuing, if the water from the New River had been sufficient."[29]

By 1745, the LBWW had had more than four decades under new owners who had been willing to invest, particularly in its pumping engines. In that year its surveyor and secretary, Samuel Hearne, produced an engineering report for the company's directors that described the state of its works in extensive detail, from its great waterwheels down to the location of all its valves in the city's streets.[30] Hearne had originally come to the LBWW perhaps as a collector and then later became its secretary. In 1739, however, he was appointed the surveyor, a post he felt a reluctance to take because of "the little knowledge I was sensible I had in the mechanics both in theory and practice."[31] Likely due to this sense of unease, he set himself the task of describing the entirety of the works. With his other duties still pressing, it took him until 1745 to finish the report. The report was roughly divided into to four parts each describing one aspect of the works: the pumping mechanisms, the water tower, the pipe network, and the employees and operations. This description is unique in the archives of the London water companies before 1800 for its scope. While substantial material from water companies exists from before 1800, Hearne's report from 1745 provides details about the company's technology and operations that can only be pieced together in the other companies. Even the records of the New River Company do not contain a technical description of such detail from this time. The report reveals a few important features. First, technology was evolving. Just as at the New River, from 1700 the LBWW's works were changing technically. This is most apparent in the waterwheels but also in the pipe network, which featured experimentation to make its supply more robust. Second, the LBWW was still very much bound to the bridge. The Broken Wharf site was no longer used to pump water. Third, the quantity of water supplied was abundant. Although wastage was enormous, the LBWW seemed little concerned with preserving supply. The problems lay mostly with actually delivering the water to houses rather than in getting enough out of the river.

The Engines

The pumping mechanism within the three arches consisted of five waterwheels driving sixteen pumps. The wheels turned in the flow, but since the tides affected the river flow strongly, their speed varied according to the time of day, and even stopped for forty-five minutes when the tides were turning at the high and low points, shutting the water supply completely. The direction of rotation of the waterwheels then reversed with the direction of the river's flow.[32] It was at these still moments that the company's employees cleaned and repaired the wheels. The mechanisms under the first two arches had been made almost exclusively of wood from their first construction in 1582 up to 1700, with the exception of brass axel sockets and pistons.[33] It was this pumping technology that Morris had introduced from the continent.[34] After the reorganization, more components, such as the cranks and cylinders, were made of cast iron, an innovation Sorocold had introduced.[35] The new material allowed the motion of those components to be more regular. Until the 1700s, iron was too scarce and expensive to be used for waterwheel construction except for a couple of components, notably the iron hoops used to reinforce the axels, and the gudgeons (the spindles inserted into the ends of the axel).[36] The increase in the manufacture of iron goods through the late seventeenth and especially in the eighteenth centuries had made these sorts of components more readily available.[37] The broader changes in the English economy, with the introduction of the charcoal smelting of iron in the early eighteenth century and late new puddling techniques to refine production, dramatically increased the volume of iron produced. This had the effect of shifting the material that waterwheels were made of from wood to iron. The wider use of iron in waterwheel construction started after 1750.[38] In 1817 the LBWW replaced its waterwheel under the fifth arch with one made entirely of iron.[39] By 1850 wheels made entirely of iron were standard.

The waterwheel in the first arch (called the "two ring wheel") was rebuilt in 1728, when the channel through it was deepened to try to increase the flow. It was 22 feet 2 inches (6.76 m) in diameter. Unfortunately, there was a sewer emptying into the Thames immediately above the bridge, sending foul water through the wheel and into the pumps. At low tide, the water was so "muddy" that the wheel was not used lest the filth jam the valves and pistons. The pistons did not draw water directly under this arch. Rather, a well was dug into the starling (arch pier). Water from the second arch was passed into this well through a pipe, and the suction pipes inserted into it. The functioning of this wheel and pump mechanism was described by Henry Beighton in 1731 (fig. 5.5).[40]

The waterwheel (the "three ring wheel") within the second arch had been built in 1732. The two wheels that stood there earlier had been removed when the lock had been widened and deepened in 1717, but they had not been

Figure 5.5. Waterwheel in fourth arch of London Bridge. Henry Beighton, "A Description of the Water-Works at London-Bridge," *Philosophical Transactions* (1731), p. 12

immediately replaced.[41] The new one at 24 feet 8 inches (7.5 m) in diameter was much larger than the two earlier ones. These used to drive only three pumping engines between them at a slow rate, while the newer wheel drove four. The current through this arch was much greater than the first, and the lock deeper and wider, so that the quantity of water it pumped was proportionately greater.[42]

The fourth arch had the largest mechanism within it, including three waterwheels and ten pistons. The first wheel in the current (the "upper wheel"), was located on the bridge's west side and was the largest of all. It had been rebuilt in 1742 with a diameter of 24 feet (7.3 m). At that time, the number of cogs on the wallow wheel was cut from twenty-five to twenty, as in other wheels in the works, in order to reduce the pumping frequency but increase the force. This had decreased the friction in the pistons as well. The LBWW used to receive complaints from its turncocks in Fleet Street, which was supplied from this wheel, but after the alterations, these diminished notably.[43]

The second wheel in the arch ("middle wheel") was rebuilt in 1726 and drove two engines. It was the smallest wheel at 19 feet (5.8 m) in diameter, and unlike the other wheels described so far, this one had been designed to rise and fall with the tides to ensure smoother functioning. Hearne observed that unfortunately this required vigilance to work well, but "the watchman is lyable to sleep in the nights," and his duty was poorly done. Although it could have been effective, the wheel's vertical mobility had become a liability. It was very expensive because the mechanism needed to raise and lower the wheel was costly to construct and required many repairs to fix the defects stemming from the irregular motion of the wheel and the wobbles it induced. Hearne judged that it brought little benefit.[44]

The final wheel ("lower wheel") was the last in the fourth arch. It was large at 22 feet (6.7 m) in diameter but less effective than the two upstream wheels. It had been rebuilt in 1734 but was too large, sitting too deep in the water. Though it was meant to rise with the tide as well, it did so too slowly. The result was that it threw water back into itself, slowing the mechanism as well as the current through the arch, thereby hindering the other two upstream wheels. At low tide, the suction heads were too high in the water, and often drew air into the pipes. This air was then forced into the main network, causing serious obstructions and even bursting mains. It cost four hundred pounds per year to repair.[45]

Hearne estimated the total volume pumped per day was 8,884 tuns (8.48 ML), of which one-sixth was lost to wastage in the engines, making the effective net volume delivered out of the pistons to be 7,404 tuns (7.07 ML), or about 1 tun (955 L) per customer. This gross volume pumped is approximately the same that Mylne recorded about thirty years later (table 5.2). Much of this water was wasted, either by the "fountaine of water running over the top of the waterhouse tower into the Thames again" or by "the rivers of water constantly running in the streets occasion'd by the overflowing of the inhabitants cesterns where they have noe ball-cocks." In addition, many of the pipes in the streets leaked into sewers and drains.[46] Hearne thought there were twenty to thirty leaks in the mains at any given time, either from decay or damage or from human error.[47] The total loss to waste was no more than one-eighth of what was pumped, meaning about 0.86 tuns (824 L) per customer was delivered. This is still a very large amount, even accounting for large customers such as breweries, which used around 15 percent of the water.[48] In all likelihood, Hearne had underestimated the loss due to leakage in the pipe network, and the actual figure would have been lower. In any case, the supply was generally deemed good. There had been no complaints received in the three months previous to the report.[49]

The basic functioning of the pumping mechanisms in use at the LBWW was that each waterwheel turned a large gear (the spur or cog wheel) attached to its end through a common axis. The spur wheel in turn drove another gear with fewer teeth (the wallow wheel), meaning it rotated more slowly. The wallow wheel then turned a crankshaft, which transformed the rotational motion of the wheels into the up-and-down vertical motion of four rods. The rods drove the pistons in the pumping engine through levers. The pumps were vacuum piston pumps, meaning they raised water on an upward stroke through one-way valves. The valves then shut on the downward stroke of the piston, forcing the water out into the pipes leading to the water tower. The piston heads were connected to the cylinders by leather hoses to prevent the water from rushing out in the space between the head and the cylinder. These leather components wore out quickly and needed frequent replacement.[50]

Valve were attached to the pipes leading to the mains, and these allowed the company to slowly drain all the mains during cold weather just when the machines sat idle when the tides changed. This kept water moving in the pipes and prevented them from freezing. The pipes went to both the mains and the water tower. If the resistance in the mains was too great, the pressure was relieved through the second pipe into the water tower.[51]

The Water Tower and the Mains

The pipes coming out of the engines led directly to the mains, with all but one main supplied by two engines. Vent pipes linked to them as they passed under the water tower. These vents relieved the pressure in the main if water was not moving along them fast enough, thereby preventing them from bursting in the streets. The vent pipes fed two lead cisterns, one atop the other, at the summit of the water tower, each with a two-tun capacity. Further pipes drained these cisterns back into the mains, with the ones that usually had greater problems meeting demand receiving water first. This system in effect transferred water from mains that had too much supply from the engines to those that required more. If the cisterns overflowed, the extra water ran back into the Thames.[52] The two mains leading into the cisterns at the high point, and therefore least likely to get water from the cisterns, were connected to one another by a valve so that water could flow between them before reaching the cistern.[53]

There were nine mains in all. Of these, the Aldgate, Bishopgate, Cheapside, Bread Street, Newgate Street, and the Fleet Street linked to the upper cistern. The Gracechurch Street, Cannon Street, and Thames Street were in the lower. Hearne had reorganized the cistern system in 1741 because the Cheapside main was suffering supply problems stemming from the Fish Street hill up which the pumps had to drive the water to reach the supply zone. Many customers were leaving to go to the New River Company.

At this point in his report, Hearne discussed the question whether hydrostatic theory was of any use in designing a water supply network. It had been well known since antiquity that water reaches an equal height in any system of communicating pipes. But in a dynamic system, the situation was more complicated, as Hearne explained. Although the water tower cisterns were more than a third higher than the Fish Street hill that the Cheapside main ascended, "every proposition in the theory of mechanics doe [*sic*] not always answer in the practical part."[54] Hearne warned that people needed to take account of the impediments that arise in the actual implementation of schemes. "Tis upon this very acc[oun]t that numbers of schemes of mechanical performances have prov'd abortive to the ruin of many hundreds of persons, by running to [*sic*] hastily into them before they have duly consider'd the impediments and have had ocular demonstration."[55] In the case of the LBWW's water, there were three causes why practice did not match theory in terms of height achieved. First, the long distance between the bridge and the end of the pipes created a hindrance to the water's motion. Second, the pipes were full of obstructions, either from foreign matter such as dirt and stones or from its own shape, notably elbows and joints. Finally, the pipes were simply too small to conduct the quantity of water the engines pushed, and the excess water was vented through any small cracks.[56] From Hearne's point of view, contemporary scientific theory had little direct bearing in running his waterworks.[57]

Air in the mains was a serious problem for the water companies. When the LBWW was first built, the waterwheels drove one or two pistons.[58] As the size and number of mains went up, more air entered the system, and it "us'd to be the occasion that the water would not (especially in great lengths) issue out of the ends of the pipes regularly, but only alternately, and in gulps."[59] Adding a third piston helped, and when Sorocold rebuilt the wheels, there were four, which was enough to give constant flow. In an attempt to determine the effect of air on the engines and in the main, Hearne described a series of events occurring in 1741–44. In the first, as an experiment, he plugged the Fleet Street main in Thames Street, and despite the wheels turning at a good speed, no water reached the plug. When a fireplug along the main was loosened, it exploded skyward, followed by air and then balls of water, rising to a great height. The air in the mains, in other words, was able to stop the force of the engines completely. A second event took place when the turn-cock caring for the Cheapside main noted that the flow was weak, although the wheels were turning well and the water tower was overflowing. Hearne opened the stopcock on the main just next to the engines, and air "discharg'd itself & roar'd like unto thunder" with no water. After a few minutes of this, the water finally came, and regular flow was reestablished. In the final in-cident, the Bishopgate main was providing weak supply. The lead and iron pipes closest to engines feeding the main would burst repeatedly, more fre-quently than Hearne's experience suggested was normal. Hearne then drained the main of all air pockets by opening all the plugs along it and pumping it until flow was moving well. The problem disappeared thereafter.[60]

These sorts of problems arose when the engines sucked air into the pipes at low tide if there were holes at the end of the suction pipes above the water line. Air could also enter the mains by other means, such as when repairs were done on the pipes. The company's usual method for draining the air out was to force it by water, trusting that enough customers had left their service pipes open waiting for water to let the air vent. Failing this, the pipes could burst. Hearne was planning on adding stopcocks to the mains near the engine to prevent them from draining during engine repairs.[61]

Maintaining a good supply throughout all the mains, especially in extra-ordinary circumstances, was an active process involving all the turncocks as it was for the New River. The turncocks took care of two or three mains each and were responsible for opening and shutting valves along the mains to distribute water according to a schedule throughout their supply district. Should there be a fire, they would rush to the appropriate valves to send all the water possible to serve the fire engines. Turncocks on other mains, more-over, would be alerted, and they would redirect some of their own supply into the affected main. There was a series of pipes with valves linking each of the adjacent mains spread throughout the city. These links meant that, even if all the pumps serving a main were not functioning, it could still be filled with water from the other engines.[62]

Although Hearne judged "these works may justly claim the preheminence of any other water works in the world from their situation, power, constant performance, and extensive usefullness," some improvements were in order. The devices that floated wheels D and E in the fourth arch could be elimi-nated because they were too expensive to maintain and gave only marginal benefit. In the second arch, the six-inch pumps could be replaced with one of seven-inch bore to help at low tides in supplying the Cheapside main. For the mains coming from the engines, on the other hand, seven inches was too small a diameter and created too much friction against the flow. Here, eight or nine inches would greatly improve the situation and be less liable to burst under the force created by the engines. This change, however, would cost on the order of two thousand pounds. Given that the tenants were generally well served with water, he doubted the directors would agree to this until they leased the third arch of the bridge from the City, at which point the fresh pipes could be of nine-inch bore. Putting new valves on the mains now, however, would allow the water to be shared between them more easily and facilitate repairs.[63]

Additional recommendations concerned the pumps. Although the cranks were cast from a single pattern, and the piston barrels were all the same length, the strokes the pistons made were varied and irregular. This was caused by the inattentiveness of whoever had installed the components in ensuring they were installed identically, particularly the regulators and forcing rods. Hearne claimed that rebuilding the regulators would resolve this problem. A related problem concerned the forcing rods, which were also irregular in length and not properly aligned to the vertical, causing breakages and avoidable friction in the pumps.[64]

Finally, Hearne suggested replacing all the square elbows with curved ones in the main network, which would decrease the friction and ease the flow. Similarly, because the mains contracted in diameter along their length, causing avoidable friction and resistance, an "infinite advantage" would result if a uniform seven-inch diameter was employed from the engines up to the very last service pipes. Whereas the mains were the primary distribution pipes, the service pipes were the ones that connected to customers' houses. Currently, the "highest tenants would have a greater and more lasting supply of water which at present in many places is very precarious especially in neap tides and dry seasons." The mains diminished from seven- to three- or even two-inch bores in some cases. These contractions had the same effect as valves in causing friction and hindering "the water from dispersing among the services with that velocity as it is raised." The mains should have diameters greater than their service pipes, particularly if, as was almost always the case, many consumers of water drew from it.[65]

The largest part of the report was a complete description of the LBWW's pipe network. Hearne described each of the nine mains down to the location of every valve and minor service pipe. The company itself had not known the exact locations of all its pipes before this work. It had been building and modifying the system for a century and a half by the time Hearne got to work. He admitted that he almost gave up the task for its difficulty and the frustration of finding the course of pipes under the streets:

> I must confess, at first setting out, I was almost disheartened from any further attempt. From the great number of pipes beyond my apprehension as well as . . . by the multitudes of persons, coaches, and carts, constantly passing to and fro in this great citty which so retarded my progress that I was determined to lay aside my chain and have recourse to the general map of London for the measurement of the pipes. But I soon found I could not have the least dependance there for truth; for the pipes lye in several places where the map takes noe notice of and some streets they over run; and in others they do not run the whole length.[66]

Hearne was finally able to succeed by waking very early on summer mornings to do his surveys before the traffic became heavy.[67]

The LBWW did not know its own network in detail because, like the New River's, it had grown organically, without any predetermined plan. Water was being supplied as far as the pumps could drive it. The turncocks were certainly aware of the locations of all the valves as they would turn them on a daily basis, but this was evidently a job learned by apprenticeship. One turncock would train the next so that he would know the order and timing of the turnings. Pipes needed to be replaced regularly, but their location was only known vaguely in a centralized way until Hearne's survey. It is not evident what the LBWW was doing to maintain control over its assets to this point. Given the significant investment that the new owners had made in the company's work over the previous forty-five years in rebuilding waterwheels and adding arches, it is remarkable that they seemed to know so little about the situation of their infrastructure; Hearne had to discover a great deal for himself in preparing his report. It may be that the LBWW had never experienced a crisis of scale like that which the New River had faced before 1700, and had never felt the need to adopt the more sophisticated methods of audit and control like that which the New River developed after 1700, as described in chapter 4. The transition to a new kind of technological complexity requiring more sophisticated control mechanisms had not occurred with the LBWW.

The LBWW served water over an area extending to Chancery Lane in the west, half the length of Whitechapel Road in the east, and almost to Shoreditch in the north (see fig. 5.6). Some streets had many LBWW mains running

Figure 5.6. A map of the London Bridge Waterworks' mains in 1745. The location of all mains has been reconstructed from Samuel Hearne's report (1745), LMA ACC/2558/MW/C/15/102

through them. Fish Street hill to Gracechurch Street had five mains from the bridge up to Fenchurch, and many others, such as Leadenhall and Thames Street, had three for some distance. Although the mains each had their own supply zones, these overlapped, and interconnections between the supply subnetworks allowed water to be transferred from one to another should there be breakages or interruptions to supply caused by repairs.[68]

This supply area overlapped extensively with the New River's. By the 1740s, the New River was able to reach everywhere that the LBWW supplied. The LBWW's great disadvantage vis-à-vis the New River continued to be its irregular supply. The LBWW was at the mercy of the tides, whereas the New River had water available in its reservoir at all times.[69] For this reason, the City of London itself used the New River's water, not the LBWW's, to fight fires.[70]

The mains in the streets were 58,965 yards in total length, or just over 33.5 miles (53.6 km), being around 10 percent the size of the New River's. About 92 percent of this was of wood, with the rest of lead and iron. The cost of the pipes at current rates was £10,378, with another £5,921 for the waterwheels and pumps, for a full total cost of the machinery of £16,300.[71] This is equivalent to £3.1 million in 2015. This was a very small figure compared to the £36,000 paid for the works in 1701, and the £70,000 capital the Chelsea Waterworks raised and spent between 1720 and 1740. The difference stemmed in part because the sum did not include labor costs to dig trenches, install all the pipes, and repair pavement. As a point of comparison, the first major canal constructed in England was the twenty-nine-kilometer Bridgewater Canal near Manchester, which cost £168,000 to build in 1761.[72] The River Kennet Navigation, which got a private act in 1715, raised about £44,000 by 1730 for improvements over eighteen kilometers. The Weaver River improvement cost £18,000 between 1721 and 1731.[73]

The Employees and Operations

The company employed nineteen people, although it contracted for others regularly to lay pipes and pave the streets. The nineteen included a secretary who kept the books and communicated with the company's directors. The surveyor was also responsible for keeping the works in good repair and, as such, supervised the paviours, who made necessary repairs and lay new pipes in the streets. The surveyor also managed the turncocks, as well as the millwrights at the company's works who manufactured new parts as needed. A storekeeper kept the inventory of all parts.[74]

There were also four collectors who dealt with the LBWW's tenants. They tried to sign up new customers if possible and kept the log books of who should be making payments on a quarterly basis. They also verified when houses became vacant so that no water was drawn there by thieves. Finally, they assured that the turncocks were keeping to the established schedule and that the customers were properly served with water. Should there be a problem, they would try to find the turncocks somewhere in their areas (or walks) and work with paviours to remedy defects in the pipes. In total, the LBWW had 7,320 tenants. Since some of these were landlords who owned a number of houses, it served more than 8,000 houses in all.[75]

Like the New River, the LBWW used turncocks to open and shut valves so as to distribute the water to houses. It employed five of them, who checked the complaints books at the main office at high and low water when the wheels were not turning. Tenants who had complaints would go to the main office on London Bridge to speak to the doorkeeper. The turncocks tried to determine what substance there was to the complaint and fix it if possible. They were also supposed to report back on all faults that had been fixed.

In addition to these there was a "supernumerary man" who had no set description, although he mostly acted as a replacement turncock. A doorkeeper and housekeeper both worked at the main office, as did a watchman at night. Finally, there were three millwrights who worked from 6 AM to 6 PM and kept the waterwheels and engines running. At every low tide, they would descend down to the water line where the suction pipes were to check for problems and keep the intakes clear. They also had to work on alternate Sundays to take care of the works and to be ready to come in during the night at any time of year for repairs if needed.[76]

The LBWW after 1750

The works did not change significantly from 1745 until 1761. In that year, the LBWW leased the third arch of the bridge, an empty one situated between two it already occupied.[77] In addition, in an attempt to address one of its competitive disadvantages with the New River, it built a steam engine at large expense on the north shore of the river at Broken Wharf to supply water at low tide when the waterwheels were stopped.[78] Finally, it also started to supply Southwark by laying a pipe from its waterwheels on the north end of the bridge along its length to the south side, and there supplied a zone in the immediate vicinity of the bridge. The third arch had been empty since 1718, and the LBWW had tried twice unsuccessfully shortly thereafter to lease it. Despite this new waterwheel, however, the LBWW by 1765 was running short of supply capacity, a problem it had clearly not experienced when Hearne wrote his report in 1745. Part of the problem was that the City had widened the main arch of the bridge by merging two of the central arches to improve the flow. To the extent this succeeded, it also decreased the flow through all of the LBWW's waterwheels.[79] Two other arches (the Long Entry and the Chapel locks) had been closed to compensate, but this only made the current through the main arch even stronger, creating a powerful eddy downriver that threatened the ships passing through the arch. The supply problem was made evident at a recent fire on Bishopgate Street, when "notwithstanding [that the LBWW] have had the assistance and advice of the most experienced and able artists in mechanics," it had taken over two hours to heat the steam engine sufficiently to raise water to the fire.[80]

To gain more capacity to make up for what had been lost with the merging of arches, the LBWW petitioned the City of London for a lease on the fifth arch of the bridge, but by proposing to occupy ever more of the bridge, the company was approaching the limits of what was available, given the other interests involved with the bridge; shipowners petitioned against giving up another arch to the LBWW because they feared it would make the eddy through the main arch even more dangerous. The City Council for its part was divided. Some councillors were willing to grant another lease, but others raised issues with the main running along the bridge to the south shore. It was leaking and clearly deteriorating parts of the bridge, especially the main arch in the center of the river.[81] The City finally decided to commission a number of surveyors (engineers) to report and make recommendations on the best way to deal with the matter. One idea they were to consider was that of leasing yet another arch, this time the second from the south shore, and blocking the southernmost arch to increase the current to the second one. This would allow the pipe along the length of the bridge to be removed. The questions were passed to John Smeaton, Thomas Yeoman, Robert Mylne, and James Brindley. With the exception of Mylne, who objected to new leases on the grounds that, expiring in 2082, they were too long, all the other surveyors recommended the plan for two new waterwheels.[82] Shipowners, however, still raised concerns about the speed of the current being dangerous for navigation and liable to undermine their piers around the bridge, and so the affair was sent back to the surveyors, this time Smeaton, Yeoman, Mylne, and John Wooler.[83] In response, Smeaton and Yeoman reiterated their original opinions and recommended approval for the two waterwheels, while

Mylne was joined by Wooler in arguing for keeping the flow more open and less powerful so as to diminish the wear on the bridge. The City Council voted to approve the two leases, with the sole proviso that they could be canceled by the City at any time should the condition of the bridge require it.[84] It was to be Smeaton himself who designed and built the wheel for the fifth arch. The new wheel was 32 feet (9.75 m) in diameter, much larger than the other wheels. He reduced the number and increased the size of the float boards along the wheel, as well as changing the size and gear ratios of the pumping mechanism. All of these improved its efficiency.[85] It was rebuilt and improved in 1789 and finally replaced with an iron mechanism in 1817 (fig. 5.7).[86]

In 1779 a fire that started in a building adjacent to the bridge destroyed the water tower but did not reach the waterwheels or pumps. Almost all the archives of the LBWW went up in flames, but the 1745 report escaped unscathed. With their now-deprived tenants appealing to the New River for water, the company managers considered whether to rebuild the tower or not and discovered by experimentation that they could pump water directly into the mains without a reservoir and still give a satisfactory supply.[87] They connected the steam engine to the mains as well, although it had normally been used only at low tides. The new arrangement was made possible because their engineer at the time, John Foulds (1742–1815), designed a cylinder into which all the pumps forced water, and to which all the mains were connected (fig. 5.8).[88] Foulds had come from Derbyshire to work as a millwright at the company in 1763 and would have become well acquainted with the waterworks, given the ceaseless maintenance they required. By 1776 he was the chief millwright. His design work on the new cylinder won him the directors' confidence, and he was made the company's engineer. Foulds became known as a master millwright and was important in the Master Millwrights Association. John Rennie, when he visited London in 1784, observed that "it was his duty to consult the most eminent master in London such as . . . Mr. Foulds (Engineer of the famous London Bridge Waterworks) as to the usages of the trade." In 1791 he was also appointed assistant engineer in the City of London's Surveyor's Office, leading to his involvement in helping to design the London Docks. Finally, he also consulted for the Shadwell Waterworks after 1797 in rebuilding a Boulton & Watt steam engine.[89]

Although the LBWW recovered from this setback, the company came under pressure from various sources from 1780. At this time, its total income from water rents was £9,686, and it was paying fifty shillings per year per

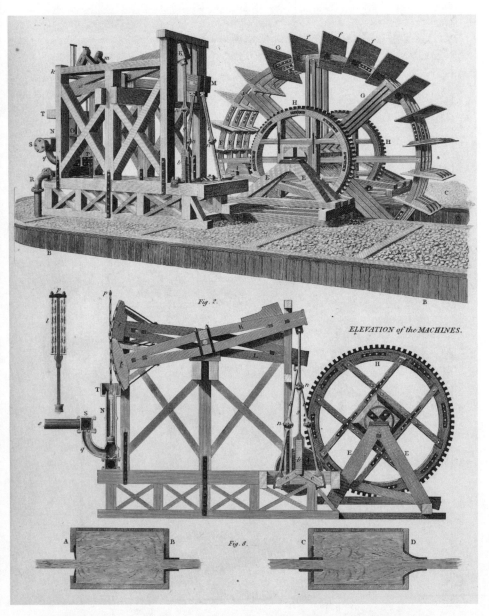

Figure 5.7. The waterwheel in the fifth arch of London Bridge in 1819. By this point most of the wheel components were made from iron. Abraham Rees, *Cyclopædia* (1819), plates vol. 4

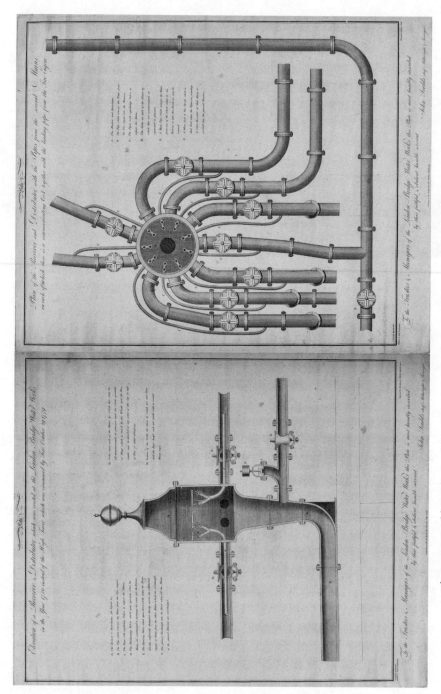

Figure 5.8. Foulds's cylinder from 1780, top and side view (drawn in 1788). Courtesy City of London

share in dividends, representing £3,750 in total.[90] The New River in that year divided £22,824, a little over six times as much. In the 1660s, the New River was paying five thousand pounds to the LBWW's one thousand pounds, so in the intervening century, the LBWW had managed approximately to keep growing alongside the New River, although not at the same pace. Their fates would soon diverge dramatically, however, because the LBWW had exhausted the potential offered by the bridge, as was made clear by a further attempt to modernize its works. Its managers had to embark on an extensive set of repairs that saw them rebuild every waterwheel they had except the one in the fourth arch. In addition, they agreed with Boulton & Watt in 1784 to replace the steam engine at Broken Wharf with one of Watt's patented engines for six hundred pounds plus fifty pounds per year thereafter.[91] The company engineer, Foulds, was responsible for all this work.[92] The LBWW managed to do all this only by exhausting all its capital and borrowing seven thousand pounds from its own treasurer James Wood.[93] Most of these works were completed by 1794, and although the new engines were more efficient than the ones they had replaced, the total income from water rents was only one thousand pounds over what it had been in 1780. Another source of pressure on the company was the changing face of the City of London itself, which was its primary area of supply. Many houses were being cleared away for commercial buildings.[94] The new Bank of England building lost the company tens of houses, as did the removal of most houses along the Thames to make way for wharves and warehouses. The East India Company was also clearing away houses to make way for its entrepôts. A final headwind for the LBWW was, as usual, the New River Company.[95] Although the agreement not to poach each other's customers under certain conditions was still in place between the two companies, the LBWW consistently lost customers to the New River when it tried to increase its rates, effectively locking in prices over decades. The only opportunity for growth came in Southwark, where the population was growing and competition was less, although even there the founding of the Lambeth Waterworks in 1785 posed a potential threat.[96] The result of all these pressures was that the managers were forced to lower the dividend to forty shillings per share in 1794, and then twenty shillings after that, before finally rising to thirty shillings in 1798 when the whole debt incurred in 1784 was finally repaid.[97] Its total dividend was less than that of the Chelsea Waterworks between 1795 and 1797, and just above in the period after, indicating that the Chelsea was finally passing the LBWW to become London's second company measured by revenues and volume of water served, pushing the LBWW aside from the place it had occupied since around 1630. By that point, however, the Shadwell Waterworks was serving more houses, but nowhere near the same volume of water or revenue. The Shadwell Waterworks had acquired the southern district supplied by the West Ham Waterworks, where the two companies overlapped in 1792, vaulting it past the Chelsea Waterworks.[98]

Conclusion

As the second largest water company in London for most of the period before 1800, and the largest before around 1630, the LBWW played an important role in the city's development. It came to provide water to around ten thousand of the houses of the rapidly expanding city, and supported industries, particularly brewing and sugar refining. In its integration into London, its capabilities and limitations were fundamentally linked to its situation on London Bridge. While the New River Company in particular was supplied by sources located at a higher elevation than the metropolis, the LBWW had to pump water up out the river and could produce only as much as its waterwheels allowed. This restricted the LBWW's capacity and effectively tied it to the City of London rather than the metropolis as a whole. As a consequence, the LBWW did not experience great growth as the city expanded through the eighteenth century, excepting to a small extent in Southwark. The LBWW suffered further as the City itself become more commercial and less residential.

The LBWW's evolution, moreover, was intertwined with the bridge's. When arches were opened up to increase the flow through the bridge, it decreased the current on the LBWW's wheels. The LBWW had to constantly seek more capacity, both by leasing more arches from the City and by repeatedly rebuilding its machines. Moreover, the LBWW's tight connection with the bridge meant that as its own mains extended deep into the City, the bridge was deeply integrated into the city's physical fabric. Over the long term, however, the metropolis outgrew the bridge, which had been adequate for hundreds of years. As it deteriorated, the City finally decided to remove it, and the LBWW went with it. The LBWW link with the City was also political. Because the City had authority over the bridge, the LBWW relied on the City's favor, especially when it was expanding. The City's generosity toward it, manifested through low rents and even loans, allowed the LBWW to be established. Moreover, over the course of the eighteenth century, the City did not clear the LBWW off the bridge even as other buildings were removed because the City relied on LBWW water.

The LBWW's history reflected evolving technology, notably improving waterwheel technology and the increasing availability of iron. As it rebuilt its waterwheels repeatedly over the years, the company sought to increase their efficiency, occasionally relying on well-known engineers, notably Sorocold and Smeaton. While not every new wheel marked a step up in efficiency, enough improvements were made to allow the LBWW to stay relatively competitive with New River despite its more expensive water, and the problems associated with being on London Bridge, including foul water close to the river banks and changing current strength as the arches were widened. The advancement of the iron industry also had its effects, as more and more components of waterwheels were made from iron. Finally, after 1800 iron was used exclusively, but this came too late for LBWW. Iron was also used for new components such as Foulds's distributer, which could contain higher pressures without needing a water tower. These shifts show how changes in the wider British economy were having effects on the secondary companies as well. These did not amount to a revolutionary change before 1800. Rather, there were many small incremental changes that nevertheless indicated a regular, or at least periodic, willingness to innovate.

Finally, the smaller companies did not have the same complexity as the New River, and the difference was not just of degree. The LBWW was the next largest company, but as Hearne's report of 1745 indicated, it had not faced the issues that the New River had confronted before 1700 as its growth outpaced its technological and business management systems. The New River had adopted new technological and management controls to stabilize itself following its period of difficulty. The LBWW by contrast had no centralized information on the situation of its works, and yet it managed to remain profitable and even expand. It took Hearne's many weeks of labor to map its network. This suggests the LBWW's relaxed and informal means of control to that time were adequate for running its relatively small business.

CHAPTER SIX

Consumption

The amount of water English people employ is inconceivable, especially for the cleansing of their houses. Though they are not slaves to cleanliness, like the Dutch, still they are very remarkable for this virtue. Not a week passes by but well kept houses are washed twice in the seven days, and that from top to bottom; and even every morning most kitchens, staircase, and entrance are scrubbed. All furniture, and especially all kitchen utensils, are kept with the greatest cleanliness. Even the large hammers and the locks on the door are rubbed and shine brightly. Would you believe it, though water is to be had in abundance in London, and of fairly good quality, absolutely none is drunk? The lower classes, even the paupers, do not know what it is to quench their thirst with water. In this country nothing but beer is drunk, and it is made in several qualities.

César de Saussure, *A Foreign View of England*, 1726

Observations on the hearty consumption of water in London were not uncommon in the eighteenth century.[1] Impressed by the scale of the infrastructure, visitors and locals commented on the relatively abundant supply available but remarked somewhat less frequently on how the water was consumed. The story of the growth of the water industry was, however, as much one of supply as of demand. If the water companies were capable of supplying water, there also had to be the demand for it. The history of London's water consisted of more than improving technology and entrepreneurship. Rather, the consumers of water were ready and willing to take the water in sufficient numbers to sustain the growth of these water companies, and they are also part of the story.

At the most basic level, the consumers of London's water had to be both numerous and wealthy enough to buy the water, as well as possessing the willingness to do so through the water networks, rather than relying on water carriers, wells, and other sources. These other sources were, of course, present and important for much of London's population into the nineteenth century. But for a majority of Londoners at some point in the eighteenth century, water companies had become their source of water. The growth of water is at one level that of early modern urbanization and especially London's exceptional place in it as it became the largest city in Europe. The water companies were inflated by London's population bubble. This, however, was not enough. London was, after all, not an order of magnitude larger than Paris, which did not develop such a system until the nineteenth century. Piped water supply, while not exactly a luxury commodity, was at least a service accessible only to people above the level of a laborer, although this gradually changed over the eighteenth century. Nominal prices hardly budged, even decreasing over that period, and the service was much more accessible to poorer people by 1800. It was, however, London's people of middling wealth and above that sustained the water industry before then. The city's wealth was a foundation for the industry. Finally, not only did people have the means to pay for the service, but they were willing to do so. Historians have posited various shifts in the willingness of consumers to buy more during this time, and clearly how they consumed water was among the choices they made.

Not only households sustained London's water industry. Industrial users, most notably brewers, were by far the largest consumers of water. London's alcohol industry in effect extended the reach of the water companies further, well into the poorer homes that could not afford water connections. Further nonlucrative uses for water, but ones that featured importantly in political discussions over the water, were firefighting and, to a lesser extent, watering roads and cleansing ditches.

Supplying Houses

Although fire ravaged the archives of the water companies on various occasions, enough evidence exists to give a good idea of consumption of water in London. Data on the number of customers served by the three largest companies (fig. 6.1) and on the number of houses connected (fig. 6.2) has been drawn from many sources, some archival and some printed, with varying degrees of reliability. Interpolations have also been required. Appendix A contains a more detailed description of the treatment of the data.

After an initial burst, the New River grew slowly from its creation to approximately 1650 (fig. 6.1). Data are unavailable from the period of the Civil Wars, but after around 1650, the customer base expanded rapidly, reaching over 9 percent annually to 1670 when statistics become available again. Its growth rate from 1670 to 1683 averaged more than 4 percent per year but was not even within this period. There was a dramatic step upward between 1680 and 1683, during which its rents grew more than 11 percent per year.

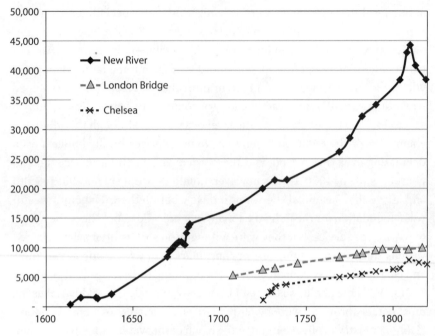

Figure 6.1. Number of customers connected

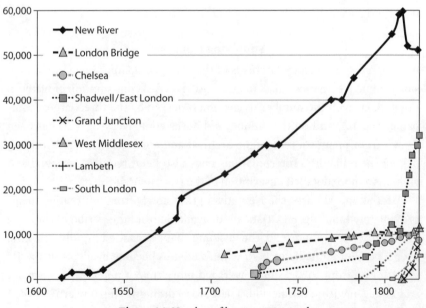

Figure 6.2. Number of houses connected

The entire period was evidently a buoyant one for the London economy. Although this expansion caused the New River problems from around 1685, once it had recovered around after 1700, its growth continued and seems to have been approximately constant until 1769 at around 1 percent per year. A building boom beginning around 1763 and ending in the 1790s caused it to tick upward to 1.3 percent.[2] Although the boom ended with the start of the French Wars, the New River's growth continued to be strong until 1804, when it exploded to 2.6 percent annually until 1810. Although it may have appeared a propitious time for the New River, a crisis soon struck it and the other water companies as technological change and new competitors took a heavy toll, effectively destroying a large part of its business model. The result was that the number of customers served then dropped 1.5 percent per year to 1820. This was a dramatic reversal for a company that had enjoyed consistent growth for almost two hundred years. The early nineteenth century crisis is described in chapter 8. The New River in 1810 was about nineteen times the size it had been in 1638. During this time, London's population had grown more than fourfold (from 300,000 to 1.3 million), indicating how dramatically the New River had penetrated into the water market in the city (fig. 6.3).

In contrast to the New River, the other companies fared somewhat less well. No data are available for the London Bridge Waterworks before the eighteenth century, but from 1708 to 1782 it almost doubled in size, a rate of growth a little less than what the New River achieved during the same period. By the end point of that time span, however, their fates were diverging dramatically. The LBWW stagnated, adding only fifteen hundred customers to 1810 (17 percent), while the New River grew by around twelve thousand (38 percent). The LBWW had reached the limits of what it could achieve based on the bridge, while the New River still evidently had ample capacity. The Chelsea Waterworks, like the LBWW, experienced slower growth than the New River after its initial spurt. It acquired customers by buying three other companies to 1732, reaching thirty-four hundred, but it did not thereafter manage to double in size even as late as 1804. Its fortunes changed at that point, and it expanded significantly over the succeeding five years, growing at 4 percent per year to 1809, only to begin losing customers in the same crisis that struck the New River at that time. From the point of view of market share in terms of number of houses served, the eighteenth century was quite constant (fig. 6.4). The New River held more than 70 percent and then dipped in the 1730s, when the Chelsea Waterworks was formed and

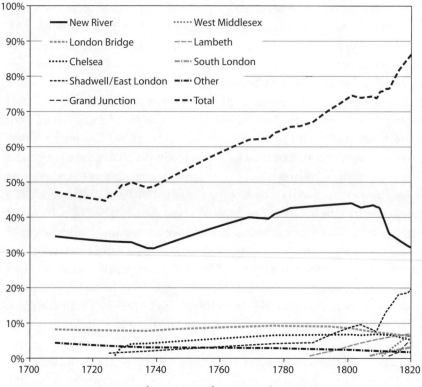

Figure 6.3. Market penetration

others expanded. Its market share held constant at around 65 percent until the end of century when new companies, first south and then north of the river, drove it down. The LBWW held a further 15 percent until around 1800. Both of these crashed after 1809, with the New River dropping to almost 35 percent by 1820.

The New River was, therefore, the most successful company up to 1800, in terms of not only absolute numbers but also growth rates. A further statistic, moreover, shows how the New River surpassed its competitors: it tended to serve more houses per customer than the other companies (fig. 6.2). Since in some cases landlords owned and paid for many houses, a single customer could represent more than one house. The data for the total number of houses connected show that the New River served roughly between 1.3 and 1.6 houses per customer, while the LBWW was around 1.1. Less data are available for the Chelsea Waterworks until very late in the eighteenth century, when it seems to have been a little higher than the LBWW at around

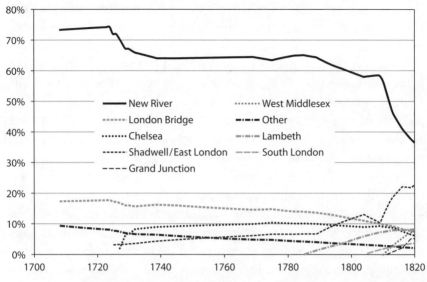

Figure 6.4. Market share in terms of number of houses

1.2. The New River's lead over the other companies was then even greater from the perspective of houses served.

The New River's dominance came to a crashing end beginning in 1806. The statistics of the new water companies formed then reveal what sort of shock they were to the London water market. The East London Waterworks was created in 1806, and by 1820 it was serving more than 32,000 houses, about a third of which came from acquiring the Shadwell Waterworks and the West Ham Waterworks. The Shadwell Waterworks itself had begun a remarkable growth spurt in the 1790s, passing both the Chelsea and the LBWW. The Grand Junction Waterworks and the West Middlesex Waterworks, while not quite attaining the same figures as the East London, still added tens of thousands of customers in very short order at the same time. Some of these customers were taken from the incumbent players, but many were previously unserved. In 15 years, the East London acquired as many houses as the New River had in its first 120. From 1806 to 1820, the three upstarts connected to more than 50,000 houses, increasing the total number of houses with water north of the Thames to more than 120,000, or an increase of more than 60 percent. Evidently, as rapidly as the pre-1800 companies had grown, they were missing out on an enormous market for piped water, one that the new companies took from under their noses with steam pumping, iron pipes, and more constant high-pressure supply. The old companies may have been technologically innovative in the eighteenth century but had entirely missed the opportunities technological changes were making possible by 1800.

What proportion of houses in London had direct water supply? Definite figures for the number of buildings in London at this time have been difficult to calculate, but estimates based on some direct counts and window and chimney tax assessments can be made for the years 1685, 1737, 1757, and 1777.[3] The national census provided firmer figures for 1801, 1811, and 1821. According to these data, the New River Company was reaching around 35 to 45 percent of buildings, with an upward drift throughout the eighteenth century (fig. 6.3). All water companies put together reached perhaps around 45 percent in 1700, with an increase thereafter as the Chelsea expanded in the 1730s to around 50 percent.[4] It increased constantly thereafter to a remarkable 70 percent by 1800, after which it held steady to 1810, when the new companies took on many new customers and about 85 percent of houses had piped water in 1820. The number of people served by the water companies was of course much larger than the number of houses since these were large households that often included extended families, in addition to servants. Given the penetration of water supply into the London market, it is clear that the New River was by 1700 no longer serving only a wealthy elite, which had been the case in its early years. Water supply reached well into the middle class of the metropolis. The dependence on water companies was such that a significant failure in a company's supply could cause distress, such as in Southwark in 1758. London Bridge was being repaired and a temporary wooden vault burned down, cutting the water pipes running along the bridge to the south. The inhabitants of Southwark were "reduced to very great distress over want of water."[5]

The reach of the water companies over the course of the eighteenth century from 45 to 70 percent of houses indicates that water became much more affordable over the long term. Wage statistics bear this observation out. The price of water held roughly constant in nominal terms from 1600 to 1800. The New River's earliest rental rate books indicate that most people paid about twenty-five shillings per year for water, a sum similar to what is recorded at around 1700. Prices drifted downward, and the LBWW charged twenty shillings per connection annually according to its agreement with the City when it leased new arches. By 1800, most water companies were charging twenty shillings per customer.[6] The exact rate was determined by the size of the house and expected use. Many people tried to haggle about fees

Figure 6.5. Ratio of the cost of water connection to wages. Wage data from www
.measuringworth.com

but were only rarely successful.[7] The ratio of the cost of water to the average
worker's wages reveals the significant improvement in the affordability of a
water connection (fig. 6.5).[8] At the industry's foundation, a connection would
have cost more than 13 percent of a worker's annual income. By the Restoration, this had decreased to 10 percent. It thereafter declined only slowly
to 8 percent over the following one hundred years. A marked decline began
in the 1750, when it decreased from 8 percent to 4 percent in fifty years.[9] By
1800, therefore, a water connection was within reach of skilled laborers, if
they had a sufficiently regular income to take a contract. The dramatic expansion of the Shadwell Waterworks from 1790, and its successor the East London Waterworks from 1810, was made possible by the cost of water connection
moving within the spending capacity of the poorer workers who inhabited the
East End, which these companies served. The wage figures used here were
for the entire country, and since London wages were on average more than
50 percent higher, a water connection was even more affordable than these
figures suggest.[10] The 30 percent of houses that did not have a connection in
1800 were naturally the poorest and would have been packed with people.
How the house figure translates into the proportion of the population served
is not known, but the segment of London's people drinking water piped to

their houses by a company in 1800 was evidently significantly less than 70 percent.

Other figures give a sense of where water stood in a Londoner's basket of yearly purchases. In 1700, after the great expansion of the water industry, Peter Earle has estimated that a median merchant in the city spent around six hundred pounds per year on domestic expenses, while a tradesman spent two hundred pounds, and a small tavern keeper or shopkeeper spent one hundred pounds. The cost of a water connection at one to two pounds per year was easily affordable for these people.[11]

The quantity of water provided to each house was not measured directly. Meters were not introduced until the late nineteenth century, and so no direct measure of the water served to the average house is available. However, commentators, such as Richard Castle in his calculations on the water supply for Dublin, or Samuel Hearne is his LBWW report, frequently reckoned 1 tun (955 L) per house daily.[12] Since houses were mostly served every second day, the average daily use was half that (500 L per day). Both Lowthorp and Sorocold used approximately the half-tun figure in their calculations.[13] Furthermore, some figures were reported for the total volume of water served by the New River at various times (1723, 1726, 1780, 1786, 1808), and they give an amount per house of 0.8 to 1.3 tuns (750 L to 1250 L) over two days, corroborating the reckoning given by the other accounts.[14] This was a significant quantity when considered in a contemporary context. In modern developed countries, water usage can be around 250 liters per person per day.[15] Assuming households with eight people, the average water consumption per person would have been around 60–70 liters daily over the course of the eighteenth century. It was this volume that prompted de Saussure's impression that "one of the conveniences of London is that everyone can have an abundance of water."[16] In upper-class homes, the water would also have been shared with thirsty horses kept in stables, and in consequence the fees charged by the water companies were higher when stables were part of the household. Not all companies supplied the same quantity, however. As may be expected, the Shadwell Waterworks and the West Ham Waterworks in the East End in 1808 supplied less, averaging one-quarter tun per house over two days.[17] Both of these companies had a very poor reputation for supply. People complained that they sometimes went two or three weeks without water.[18]

In homes, the water was used for a variety of purposes, including for drinking, laundry, washing, preparing foods, watering stables, bathing, and

TABLE 6.1.
New River Company Summary, 1769

Total number of tenants	25,910
Total gross rent	£37,324
Average rent	£1 8s 9d
Median rent	£1 4s 0d

TABLE 6.2.
Income by Yearly Rental Contract Amount, New River Company

Yearly Rent	Count	Portion of Total (%)	Total Collected	Portion of Total (%)
£1 4s 0d	7,719	29.6	£9,262 16s 0d	24.9
£1 0s 0d	4,540	17.4	£4,540 0s 0d	12.2
£1 10s 0d	1,644	6.3	£2,466 0s 0d	6.6
£1 3s 0d	1,505	5.8	£1,730 15s 0d	4.7
£0 15s 0d	1,196	4.6	£717 12s 0d	1.9
£1 6s 0d	1,162	4.5	£871 10s 0d	2.3
£2 0s 0d	1,130	4.3	£1,469 0s 0d	4.0
£0 16s 0d	962	3.7	£1,924 0s 0d	5.2
£0 10s 0d	795	3.0	£636 0s 0d	1.7

even ornamentation in gardens of the wealthiest houses. It is difficult to know what the relative importance of these various uses was and how they changed over time. As described in chapter 7, from being rare in the early eighteenth century, daily bathing at home became commonplace in middle- and upper-class homes by the end of the century. Water closets were also becoming more common. Although they had existed earlier, Joseph Bramah patented a flush water closet in 1778 that proved to be practicable enough for more common use. He sold thousands of them by 1800, and they were widespread in London in middle-class districts by the 1830s.[19] Home laundry was becoming more mechanized as well, with a number of new washing machines appearing from the mid-1700s onward, including a "Yorkshire Maiden" advertised in the *Gentleman's Magazine* in 1752.[20]

The most significant pre-1800 customer data set available from the London water companies dates to 1769 when the New River Company's head office burned down with most of its records perishing in the blaze. In order to consolidate their operations in the wake of this disaster, the directors compiled a complete list of all their tenants. The statistics are listed in tables 6.1, 6.2, and 6.3.

Although the New River Company had customers of many sizes, it made most its revenues from small customers. Figure 6.6, which shows the running

TABLE 6.3.
Largest New River Customers

Customer	Street	Rent	Volume of Water (tuns per day)	Activity
Gyfford & Co.	Castle Street	£90 0s 0d	20.55	Brewer
Truman & Baker	Brick Lane	£84 0s 0d	19.18	Brewer
Samuel Whitbread	Chiswell Street	£84 0s 0d	19.18	Brewer
Calvert & Co.	Red Cross	£80 0s 0d	18.26	Brewer
Robert Hucks	Hyde Street	£55 0s 0d	12.56	Brewer
Chase & Cox	Great Russel Street	£51 10s 0d	11.76	Brewer
Victual office	Tower Hill	£39 15s 0d	9.08	Navy supply, brewer
Samuel Hawkins	Long Lane	£38 8s 0d	8.77	Brewer
Thornton	White Lyon Street	£38 0s 0d	8.68	Brewer
Wilks & Raw	St. John Street	£38 0s 0d	8.68	Malt distiller
St. Bartholomew's Hospital	Smithfield	£37 0s 0d	8.45	Hospital
Cokar & Co.	Old Street	£35 0s 0d	7.99	Brewer
Mason & Co.	St. Giles	£32 0s 0d	7.31	Brewer
Allen & Ambrose	Nightingale Alley	£30 0s 0d	6.85	Brewer
Dickinson & Co.	Golden Lane	£30 0s 0d	6.85	Brewer
Dickenson	St. John Street	£30 0s 0d	6.85	Brewer
Bulstrod	Air Street	£28 0s 0d	6.39	
Sankethman	Shoreditch Street	£28 0s 0d	6.39	
Somerset House	Strand	£25 12s 6d	5.85	Residence and barracks
Jones	Goswell	£25 0s 0d	5.71	
Morly	Leather Lane	£25 0s 0d	5.71	
Christ Hospital	Town Ditch	£23 0s 0d	5.25	Hospital
John Suger	Bartholomew Square	£23 0s 0d	5.25	
Sam Read	White Chapel	£22 10s 0d	5.14	Brewer
Jordan & Lefevre	Leman Street	£22 0s 0d	5.02	Brewer
Joanne Deane	Cow Cross	£22 0s 0d	5.02	
Wid Clempson	Hare Street	£20 0s 0d	4.57	Brewer
New Inn Pay & Co.	Islington	£20 0s 0d	4.57	
John Edwards	Goswell	£20 0s 0d	4.57	
Murey	Lukerpond Street	£20 0s 0d	4.57	Brewer?
Broadhead	Browns Garden	£20 0s 0d	4.57	
Starkie & Co.	Brewer	£20 0s 0d	4.57	Brewer
The Charterhouse	Charterhouse Square	£20 0s 0d	4.57	Hospital and school
Charles Dodd	Mutton Lane	£20 0s 0d	4.57	

sum per contract for both number and total rental per contract size, reveals just how important small customers were. Most of the company's tenants, about 91 percent, paid two pounds per year or less, and these represented 72 percent of its total revenue. For a three-pound cut off, the corresponding figures are 95 percent of the total count and 80 percent of the revenue. The large steps in figure 6.6 represent the most common contract amounts. The nine most common, all at two pounds or less, accounted for 84.7 percent of all customers and 64.1 percent of total income. The New River Company, and by extension the water industry in London, was therefore a phenomenon founded on small consumers, as the proportion of houses connected to the network reveals (fig. 6.2).

The willingness of consumers to take piped water that these figures demonstrate should not be assumed as self-evident or inevitable. The New River's difficulties, described in chapter 2, in its earliest years in gaining customers,

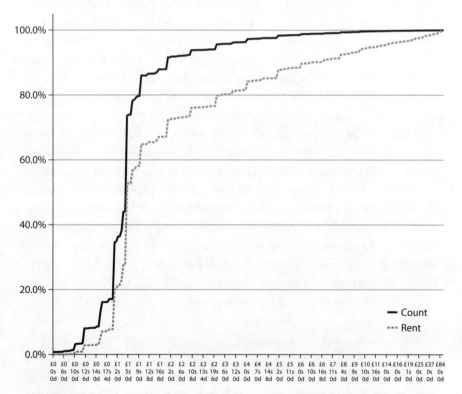

Figure 6.6. Contract size distribution. The lines represent the proportion of customers below the given yearly contract amount.

as well as the dips in their number during the 1620s, probably due to plague, indicate that factors could work against the adoption of water supply. The case of Paris in the eighteenth century is another counterpoint. The technology was available in Paris as in London, but as the history of water supply from 1600 to 1800 showed, this did not produce the same results. Many factors contributed to the difference, and they varied with time. In the early period, the Crown meddled more directly with how water was supplied in Paris, and a commercial industry was never established there like it was in London. Later in the eighteenth century, when attempts were made to set afoot water companies using a model akin to London's, consumer demand was not adequate to sustain a network.

The first waterwheel pump was erected in Paris in 1608 on the pont Neuf by Jean Lintlaer, a Flemish engineer, not long after Morris's wheel in London. Unlike in London, however, the instigator and owner was the king, Henri IV, and the pump served to supply the Louvre and Tuileries palaces, which were using about half of all water brought to the city.[21] No supply from the new pump was available for sale to houses, and the business model that Morris had brought to London never took hold in Paris. Another project sought to supply water through the Arceuil aqueduct. It was built by private contractors in agreement with the king (and then Marie de Medici after his assassination) and the city. Supplies began in 1623, but the project was required to give 60 percent of the water to the new royal Luxembourg palace, with the city and the contractors splitting what was left.[22] Although most of this made it to public fountains, there was intense lobbying on the part of nobles, religious houses, and public institutions to gain access to direct water connections, leading to the repeated cycle of diminished public supply and the revocation of privileged connections, before fresh lobbying once again harmed the public supply.[23] These connections were provided gratis if the requester was deemed to have rendered a service to the city, in default of which it was given for a fee.[24] Other large waterwheels, also supplying public fountains, were built on the pont Notre-Dame in 1673 by the city government. The city had approved its construction with the idea of offsetting its cost by selling any excess water not used for supplying fountains. The plan failed, because the water pumped was not adequate and the city kept giving away whatever water it could spare.[25] Attempts to found for-profit water companies were made in the years around 1695, but few got underway. Even those that did fared poorly; one established on La Tournelle bridge in 1695 failed in 1707 and was removed.[26]

New companies offering water for sale emerged at the end of the Ancien Régime in France.[27] In the mid-eighteenth century, with city supplies in a dire condition, and the Notre Dame pumps in a bad state, Louis XV in 1763 issued a patent creating a company to sell water collected in boats on the river, but this had little success because it used water carriers to distribute to paying customers.[28] In the meantime, water carriers drawing from the Seine became even more important for the city.[29] After many years of debating various proposals, both public and private, the city finally authorized the creation of a larger company, the Compagnie Royale des Eaux de Paris, established by Jacques-Constantin and Auguste-Charles Périer in 1778. It offered the London model of domestic water service through a pipe network fed by a steam engine (purchased from Boulton & Watt) pumping water from the Seine, but the inhabitants of Paris were reluctant to pay for the connections. The company also sold water from public fountains. It expended all its initial capital within a few years and had to issue more stock repeatedly. After struggling for a few years, the company finally managed to turn a profit in 1786, at which point the comte de Mirabeau estimated that it had only 617 customers with piped water. Moreover, the number of customers with direct connections was decreasing, while those taking from public fountains increased.[30] The company finally failed in 1788 and was largely taken over by the city.[31] A competing enterprise that sold water brought to the city from the Yvette River via aqueduct was founded in 1786, but it had no better success and failed in 1793.

The chaos of revolution and war meant little was achieved in the succeeding years so that in 1820 Paris's water supply was still much as it had been for the preceding century. Indeed, even in the 1850s, when a universal piped water system had been introduced to the city, only about sixty-three hundred buildings, or less than 20 percent, had connections. London, by contrast, had passed the 20 percent point long before 1700, probably around 1660. In Paris, most people preferred to get water from small public fountains, which were found on almost every street in the 1850s.[32] It was this preference for fountain water, which lasted well into the nineteenth century, that had hampered the growth of the Parisian companies before 1800. Nor was this preference unique to Paris. As Chris Hamlin has observed, in early modern Europe "it would appear that the piping of water into individual homes was not felt to be important." Only in the nineteenth century did it really take hold.[33] London, it seems, was exceptional. More research, however, is needed to investigate how the availability of technology was connected with consumer demand in other cities before 1800.

London's consumers proved to be eager for water, sufficiently so that they sustained the industry's constant growth from sometime in the mid-seventeenth century. The willingness of masses of Londoners to consume water in this way indicates, at least initially, a fresh attitude toward the consumption of the most quotidian of consumables: water. As the case of Paris shows, availability of piped water was not sufficient. The consumers also had to want it delivered through pipes. Changes in patterns of consumption are naturally common, but historians have argued that sometimes more deeply rooted societal changes in consumption have occurred, giving rise over time to the intensely consumerist societies of the twentieth century in the West. Indeed, Jan de Vries has pointed out that historians have made claims for five "consumer revolutions" between the Renaissance and the twentieth century.[34] For de Vries, however, the crucial consumer revolution for rise of the culture of mass consumption began in seventeenth-century Holland and spread to England in the late seventeenth century. This was certainly the most important among the putative consumer revolutions for the water industry, coinciding with the rapid takeoff of demand.

The revolution originated in Holland when innovative consumer behavior emerged so that, in de Vries's words, "for the first time on such a scale and on so enduring a basis, we find a society in which the potential to purchase luxuries and novelties extended well beyond a small, traditional elite and where the acquired goods served to fashion material cultures that cannot be understood simply in terms of emulation."[35] Uniquely in Europe to that point, a large proportion of a society was able to choose how to spend its wealth and which goods to buy and use. For de Vries, this "new luxury" revolution was not an extension of the old aristocratic culture of luxury. Rather, it differed in that it was more directed to the home, to the interior, as opposed to the ostentatious luxury of Renaissance culture. The aim was more comfort than the indulgence of refined taste, domestic rather than courtly. The interests of new consumers were in acquiring not unique valuable artefacts but "products capable of multiplication," such as Delft dishware, cabinets, furniture, books, pipes, and clocks, as well as a broader range of shorter-lived goods and consumables, including more textile fabrics and cloths, sugar, tobacco, coffee, tea, fine grain breads, and spirits.[36]

In order to acquire all these goods, people modified their work behavior by working more hours so as to be able to earn and spend more. This, de Vries argues, was an "industrious revolution" that linked patterns of both work and consumption. It was not only an increase in wages that prompted this. Indeed, hourly wages did not rise substantially. Rather, people were willing to work longer in order to change what they consumed, prompting a significant change in the basic "consumption bundle."[37] Furthermore, de Vries claims that the industrious revolution had been founded in two different but overlapping demographics. The first was the middle classes, who earned more and typically could afford servants. They concentrated their increased consumption especially in the domestic setting, being interested in furthering sociability, comfort, and respectability. The second demographic involved a more plebeian sort, the type of worker who often worked as a servant rather than employing them. Such workers could not afford the same domestic consumption as people farther up the income ladder but still enjoyed a wider variety in their food palette, tending in addition to find social interactions outside the house, such as in taverns.

The Dutch never theorized about their new luxuries, and contemporary discourse exploring and justifying them remained underdeveloped. The new Dutch culture of consumption spread to England, especially in London, in late seventeenth century, and it was there that it received a fuller theoretical elaboration.[38] Already in the early seventeenth century, there were attempts to demoralize luxury consumption by rejecting charges that it was wasteful and immoral. These attempts were successful to the degree that, although people still argued for sumptuary laws to control luxury purchasing in the early seventeenth century, these grew less common, and such bills largely disappeared from Parliament after 1640. Although the controversy about whether luxury was morally dubious hardly disappeared, over time luxuries received new, less morally laden labels such as "excellent" and "delicate."[39] In the 1670s and 1680s, authors such as Nicholas Barbon and John Houghton energetically defended consumers and consumption, arguing that the appetite for material goods and comforts promoted the public good as wealth circulated throughout the entire country. For Barbon, cities in particular were the crux of this happy situation, securing a path to prosperity and peace as seats of consumption. Barbon set an important precedent in arguing that the growth of London, which was often seen as disordered and unrestrained, was, on the contrary, beneficial for the national economy.[40]

The basis of this consumer revolution in England and London in particular was not merely rhetorical but was founded on relative economic changes and growing prosperity. Over the course of the seventeenth century, national wealth approximately doubled. Furthermore, the population was becoming increasingly urbanized. In 1525 only 5.25 percent of England's people lived in urban areas (towns of more than five thousand inhabitants). By 1600, this had reached 8.25 percent, 13.5 percent in 1670, and 17.0 percent in 1700. Furthermore, London's share of this urbanization was expanding, with up to 11.5 percent of the nation's people in 1700 from 2.25 percent in 1520.[41] The proportion of the country's workforce devoted to agriculture decreased from around 58 percent between 1500 and 1600 to 39 percent in 1700.[42] Urban workers were also earning more money, due to working longer hours rather than benefiting from rising wages, after a long period of stagnation and even decline from the end of the Middle Ages to 1650. From a level comparable to other European cities such as Vienna and Florence around 1575, workers' incomes in London had doubled by 1700 relative to these other cities, which largely held constant. Although Amsterdam's wage earners enjoyed similar growth over this time, the two cities diverged thereafter, as London's wages continued to grow while Amsterdam's stagnated.[43] The growth in wage figures is also borne out by data on the expansion of economic output. The gross domestic product of England has been estimated to have doubled between 1600 and 1700. Due to population increase, however, real GDP growth per capita before 1650 was negligible, with the figures from 1400 almost the same as 1650. Most of the growth took place after 1650, during which time it increased from £8.85 to £12.68 per capita in 1700, based on 1700 prices. It continued to grow thereafter, albeit more slowly.[44] The period after 1650 was, therefore, a particularly good one for the English economy, which, when coupled with the changing attitudes of the consumer revolution, buoyed consumption, including of water supplied in a new way.

Over the eighteenth century, the consumer revolution spread from London to other areas of the country, and farther down the income scale, so that by around 1770 laborers were adopting some of the consumption habits that had first appeared a century earlier in middle-class households. This was a second consumer revolution affecting the water industry. As Neil McKendrick has argued, in the eighteenth century people on a wider social spectrum bought things more than ever before. Increased employment of women and children raised family incomes, and retailing became more sophisticated, with advertising and marketing playing a larger role. The expanding factories of the industrial revolution employed more people, and produced the goods that supplied some of this consumer revolution.[45] Furthermore, working hours continued to increase, raising net incomes.[46] The constantly growing market penetration of the water industry throughout the eighteenth century and then its expansion in the poorer East End after 1790 show the effects of the widening of consumption on less wealthy economic groups.

Economic thinkers, too, continued to try to account for origin and nature of consumption in positive ways, as Barbon and others had in the late seventeenth century. These new philosophies embraced modern consumption as fundamentally good, in contrast with the elitist old luxury. They further believed that high wages sustained this new consumption, allowing more scope for people to be industrious and thereby enjoy the pecuniary and material fruits of labor. David Hume, for example, argued that a consumer society pursuing wealth and consumption was a way to redirect human passions away from making war toward more benign activities. He further claimed that an industrious society interested in luxuries could improve itself and produce knowledge through the refinement of the techniques of production and of designs, prompting further development in science and the arts. Adam Smith argued that "consumption is the sole end and purpose of all production." He was not, however, simply advocating old luxury style consumption but rather one rooted in individuals seeking their own good prudently in the long run, practicing self-denial and eschewing immediate gratification for measured consumption over time.[47]

Brewers and Other Large Users

As important as small consumers were, large consumers still featured importantly among the New River's customers in its 1769 audit. The largest consumers were mostly brewers (table 6.3, fig. 6.7), and in contrast to the smaller users, they paid for their water according to volume consumed. This was measured not by meter but according to the size of their cisterns and the flow rates allowed into them. In the late eighteenth century, the New River charged 30 shillings per 1,000 barrels (2.88 pence per tun).[48] This was up from the 1730s, when it charged 20 shillings per 1,000 barrels.[49] In 1769 the top consumer was Gyfford & Co., a brewer in Wood Yard off Castle Street (later Shelton Street) in Long Acre. The brewery had been substantially expanded in the 1740s and was owned by partners who included many from the Gyfford family, which had a history in the industry. Other partners were Peter Hammond, from a Southwark brewing family, and Henry Evans, a hops merchant.[50] Just a bit below Gyfford was the Truman & Baker, or the Black Eagle Brewery on Brick Lane in Spitalfields. The brewery was the third largest in London, producing 60,000 barrels (7,500 tuns) of beer per year.[51] Of equal size was Samuel Whitbread on Chiswell Street north of Moorgate. The firm produced porter and pale and amber beers and had locations around the city. It would become the largest brewer in London by 1784.[52] Slightly smaller than these was Calvert & Co. on Red Cross Street.[53] These four brewers were the pinnacle of the London brewing industry, which by the mid-eighteenth century was dominated by a few high-volume, highly capitalized brewer-ies. In 1778 the Board of Excise identified six brewers that merited being placed in the class of capital houses: Whitbread, Thrale, Truman, Sir Wil-liam Calvert, Felix Calvert, and Hammond.[54] Of these, all but Thrale was a New River customer, and this one exception was located in Southwark, away from the company's pipes.[55] Thrale owned its own waterworks, the Bank End Company, taken over by the Borough Waterworks in 1771 (see chapter 5).

The prevalence of brewers extended beyond the uppermost peak of con-sumers. Among the thirty-four water consumers paying more than twenty pounds per year, at least nineteen were brewers, and fourteen of the top six-teen. Among the nonbrewing consumers whose identity can be determined were three hospitals, the Royal Navy's victualing office, and Somerset House, which was then used as a residence and barracks. Even the victualing yard likely used the water for brewing. The remaining customers have not been identified. These largest users consumed about 275 tuns of water per day (about 263,000 L), or about what 550 houses used. Since most beer produced by these brewers was consumed in the city, many people who did not have water connections to their houses were nevertheless drinking company water in their beer.

The prominence of brewers on this list reflects the state of brewing indus-try in the mid-eighteenth century and the significant change it had under-gone in the preceding decades.[56] In the seventeenth century, most beer was brewed by small brewers that sold directly into a retail market, either through public houses or to domestic consumers. There were also some Common Brewers, as those who sold wholesale to public houses and to consumers were known. Although their numbers grew to just under two hundred by 1700, the total volume they brewed relative to the whole market was still small. Important changes came in the 1720s when porter was first brewed. Demand for this strong black bitter beer was robust, not least because it was relatively cheap. The low price of porter was made possible by efficiencies of scale in production, something for which it was especially suitable. Porter

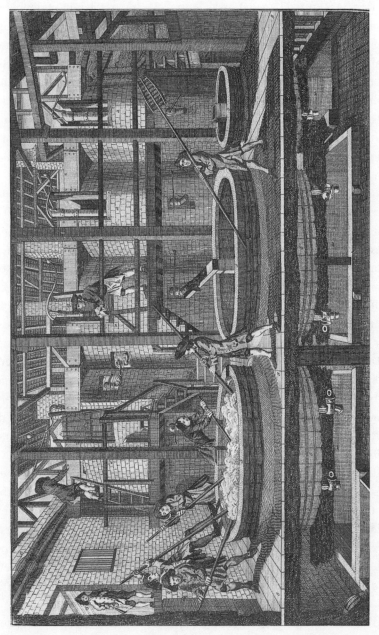

Figure 6.7. A brewhouse. T. H. Croker, *The Complete Dictionary of Arts and Sciences* (1764–66), vol. 1, pl. XXIV

was less sensitive to changes in the fermentation process than ale, which needed to be matured in wood casks; porter was more tolerant of temperature variation during fermentation, meaning it could be brewed in large vats and for a greater part of the year, into the warmer months, than ale. Technological innovation, such as the introduction of thermometers and metal vats, facilitated the expansion in scale. In addition, soft water, which the New River provided, was more suitable for porter brewing. For these reasons, only the porter breweries expanded between 1720 and 1800, while the ale ones remained small. The situation would shift in nineteenth century with technological changes that allowed ale to enjoy the same efficiencies of scale, leading to a resurgence in its popularity.

Peter Mathias has described this change in the brewing landscape in the decades after 1720 as a metamorphosis of the London industry toward industrialized brewing with mass distribution, run by a few large brewing houses.[57] Mathias argued that the rise of the large London porter breweries was an important part of the economic and technological transformation of the English economy in the eighteenth century and should be regarded as part of the industrial revolution. The importance of the New River in this metamorphosis is underscored by the production statistics. The London porter brewers produced around 1.2 million barrels of beer (150,000 tuns, 143 ML) in 1770. The brewers listed among the largest New River consumers used around 400,000 barrels (50,000 tuns, 47.75 ML). The actual volume used by New River brewers was evidently higher because many brewers have not been identified on this list, and other smaller ones are not included. New River water sustained at least a third of London's beer production and certainly much more.

Although they did not use as much water as brewers, there were other nonresidential consumers of water, including bakers, butchers and slaughter houses, chemists, cow keepers, curriers, color manufactories and dyers, distillers, fishmongers, stables, washers, public houses, soap boilers, sugar bakers, and tripe boilers.[58] Many of the heavy users, including fishmongers, distillers, sugar bakers, and brewers, drew their water at night when the effect on local supply was less noticeable.[59] A further major consumer of water emerged during the eighteenth century: operators of steam engines. Although some could draw water from wells and rivers, many fed their engines water from water companies. For their part, the water companies found these consumers to be problematic because of their voracious appetites. The

New River even began refusing connections to users who operated steam engines.[60]

Geography of Consumption

The geography of consumption of water in London in 1770 is depicted in figures 6.8, 6.9, and 6.10. Figure 6.8 represents the total revenue the New River Company derived by unit area. The contours are drawn at six pounds per year per block, where each block is 2,500 m². To the degree that fees charged for water reflected usage, this plot shows where water use was more intense in London. Evidently, because there were no water meters, the correspondence between rental and usage is approximate; but since the company did make the effort to charge per usage according to the size of the house and, among the largest users, according to volume used, the plot does approximately correspond to actual usage patterns. The figures show unsurprisingly that consumption in most of the West End was elevated compared to the rest of London. However, the usage was so low west of New Bond Street that it does not register above the first contour with the exception of high-usage areas around Grosvenor Square and, to a lesser extent, Portman Square and Berkley Square. The cause of this drop off was that the New River had more difficulty getting its water to those areas, and the Chelsea Waterworks in particular offered strong service there. Farther east in the West End, the development estate squares continued to be centers of intense use. This is notable with St. James's Square, Golden Square, Sloan Square, Cavendish Square, Soho Square and areas to its south, Hanover Square, and Bloomsbury Square. Figure 6.9 plots contours for the average yearly price paid for water. Each contour line is drawn at ten shillings per year. Because the calculation is not weighted according to consumer density, a single isolated consumer raises the contour of the area in which it was located, while large consumers located among many small ones have relatively little effect on the plotting of the contours. The graph reveals many of the same patterns as figure 6.8. The West End squares were concentrations of high paying users, all the way to Hyde Park at the western edge. In addition, the East End has some relatively high spikes caused by large isolated breweries, or the navy victualing yard close to the Tower of London. Many other brewers were located in the north and northeast of the metropolis, causing the higher contours there. There is also a large spike around Whitehall in Westminster because of its isolation from other New River customers.

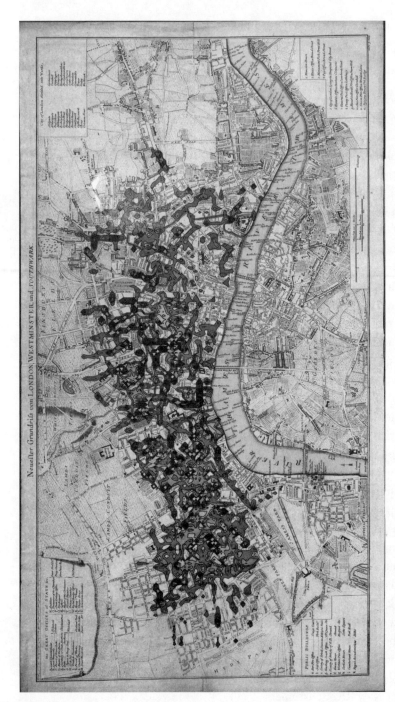

Figure 6.8. Income derived from New River customers in 1770. Lines are drawn at six pounds per 2,500 m². The West End to Swallow Street was the densest income zone. Other streets with dense concentration included the Strand. The East End, especially where the LBWW operated, was almost a dead zone for the New River. Calculated from Collectors' rents and arrears book LMA NR ACC/2558 /NR/12/001 (1770)

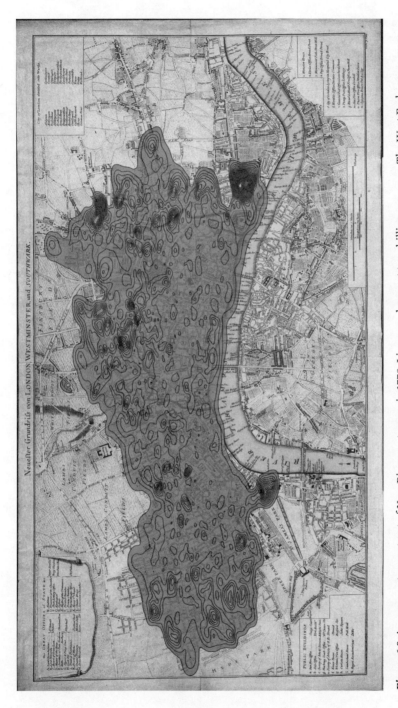

Figure 6.9. Average water contract of New River customers in 1770. Lines are drawn at ten shillings per year. The West End square were centers of higher contracts in the west, while relatively isolated brewers were located in the east and northeast fringe of the City in the East End. Calculated from Collectors' rents and arrears book LMA NR ACC/2558/NR/12/001 (1770)

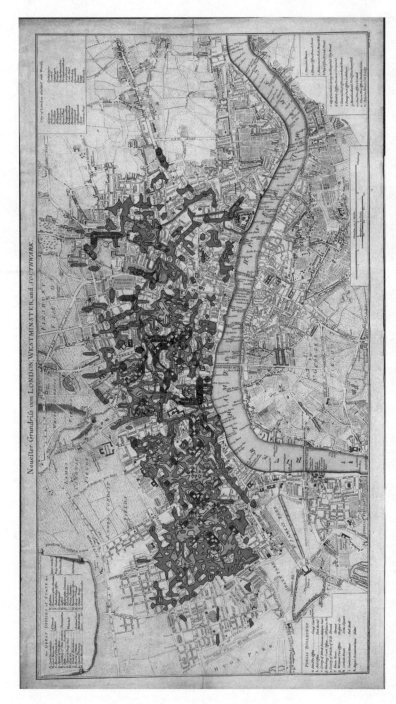

Figure 6.10. Count of New River customers in 1770. Lines are drawn at five customers per 2,500 m². Calculated from Collectors' rents and arrears book LMA NR ACC/2558/NR/12/001 (1770)

Figures 6.8, 6.9, and 6.10 reveal that the West End squares were important sources of income for the New River. Most of the squares had been built after the Restoration when speculative builders developed many aristocratic estates in the West End, beginning with Bloomsbury and St. James's. This was done in two large spurts, the first from 1660 to 1700, with another around 1720 to the 1730s. Building slackened notably thereafter, not to restart until the 1760s. All the squares had been created as London's growth swallowed the surrounding countryside, without, however, entirely eliminating its rural flavor. There were gardens as well as intermediate zones between the metropolis and the countryside. The squares also marked a new approach to the design of aristocratic houses, away from the medieval pattern of introverted and isolated buildings throughout the City, toward ones oriented to external display in a neighborhood of peers. These houses were comfortable with open air and spaces, built for easy circulation of coaches and people in the nearby streets and squares. In the late seventeenth centuries, their builders would improve the neighborhoods with amenities, such as paved walkways, drains, chapels, and markets. Frequently, builders would also provide water, meaning that new development projects were an effective means for a water company to gain access to many new houses.[61]

The earlier squares were often dominated by an especially imposing house, such as Southampton (later Bedford) House in Bloomsbury Square. This was not the case with the later group dating to the 1720s, including Hanover, Grosvenor, and Cavendish Squares. Smaller houses lined the other sides. The use and design of these squares changed over the eighteenth century. They became more socially and physically enclosed, with fencing and gates restricting access. The Restoration squares maintained their aristocratic flavor for some time, but some began to lose it as the local demographics drifted. By the 1760s, for example, Golden Square had lost its aristocratic inhabitants.[62] The downward social transformation was not so pronounced as to attenuate New River water consumption. The West End received the New River's special attention throughout most of the eighteenth century. During the London social season from May to August, when aristocratic and landed families would move to their grand houses in the capital from their country estates, more water was provided. In the late eighteenth century, for example, the New River provided water on Sundays to West End mains during the summer.[63] It also regularly increased its supply there by keeping its mains full for more hours, building more mains, and, as discussed in chapter

4, installing a large 16-inch (40.64-cm) diameter iron main from the upper pond to serve the West End in 1790.[64]

Away from the West End urban squares, spikes in demand occurred where the largest users were located, including Castle Street, Great Russell Street, Holborn, and around Somerset House, in addition to some East End locales, such as White Cross Street east of Moorfields. In addition to Holborn, a number of the major roads of the metropolis were also areas of high consumption, including along Oxford Street, the Strand and Fleet Street, Fetter Lane, Aldersgate Street, Golden Lane, Chiswell Street, Old Street, and Bishopsgate to Shorditch up to Hoxton Road. Other streets, notably in the East End, did not feature uniformly high consumption but rather had areas of more intense consumption where larger consumers were. Streets that displayed this spotty pattern included Brick Lane, Whitechapel, Tower Hill, and Nightingale. Finally, some other neighborhoods showed higher densities of water use, notably south of Lincoln's Inn, the areas around West Smithfield, the region between Gray's Inn Road and St John Street where brewers were located, and around Moorfields in the east.

Figure 6.10 maps the number of water consumers by 2,500 m² block with the contours at five customers per block. Although the overall pattern is the same as the revenue plot, the differences are also instructive. The West End squares are less important here, indicating that the high water use shown in the figure 6.9 stemmed from the palatial homes that sat around those squares. Indeed, among the 935 streets listed in the audit book, many of the squares are in the top 5 percent in terms of highest average rental per customer (see table 6.4). St. James's Square, for example, had seventeen tenants paying an average of £4 17s 6d per year. Grosvenor Square had twenty-three paying on average £4 3s 1d, while Hanover Square had twenty-six paying £3 11s 6d. In contrast to the squares where a few tenants consumed a great deal, other areas of higher consumption reflected denser packing of smaller users (see table 6.5). The region south of Soho Square, as well as along Swallow Street especially around Glasshouse Street, were busy in this way, showing no drop in density from the revenue plot to the population plot. Other dense areas included the Strand, which had been the site of grand aristocratic houses until the late seventeenth century when they were replaced by smaller houses occupied by gentry;[65] Fleet Street, which was an important shopping and market area;[66] the areas north of St. Paul's Church up Aldersgate and around West Smithfield; and Shoreditch and areas near Moorfields. Some of these

TABLE 6.4.
Streets with Highest Average Rent (>10 tenants), New River Company

Street	Number of Tenants	Mean Rent	Total Rent
St. James's Square	17	£4 17s 6.4d	£82 18s 0d
Grosvenor Square	23	£4 3s 0.5d	£95 10s 0d
Hare Street	16	£3 17s 6d	£62 0s 0d
White Hall	16	£3 15s 10.5d	£60 14s 0d
Hanover Square	26	£3 15s 1.2d	£37 11s 0d
Cavendish Square	25	£3 11s 6.5d	£93 0s 0d
George Street Great	14	£3 11s 0.5d	£88 16s 0d
Portman Square	14	£3 9s 3.4d	£48 10s 0d
Hill Street	22	£3 7s 0d	£33 10s 0d
Soho Square	25	£3 6s 5.1d	£46 10s 0d
Rupert Street	11	£3 6s 0.5d	£72 13s 0d
St. Johns Street	17	£3 6s 0d	£82 10s 0d
Lincolns Inn Square	27	£3 5s 10.9d	£36 5s 0d
Chiswell Street	67	£3 5s 0.7d	£55 6s 0d
Old Street	97	£2 19s 4.9d	£80 4s 0d
Red Cross	72	£2 16s 2.5d	£188 6s 0d
Town Ditch	14	£2 15s 7.9d	£269 19s 0d
Charles Street	30	£2 15s 6.3d	£199 17s 8d
BW Gardery	21	£2 14s 11.1d	£38 9s 0d
Clifford	13	£2 13s 9.2d	£80 13s 0d

TABLE 6.5.
Streets with Highest Total Rent, New River Company

Street	Number of Tenants	Mean Rent	Total Rent
Moorfields	266	£0 15s 5.3d	£205 7s 0d
Fleet Street	308	£0 12s 10.9d	£198 16s 0d
White Cross	209	£0 19s 0.1d	£198 12s 1d
St. John Street	228	£0 16s 9.8d	£191 14s 6d
Aldersgate Street	217	£0 17s 1.1d	£185 8s 6d
Fetter Lane	254	£0 13s 11.3d	£177 0s 4d
Strand	213	£0 13s 7.1d	£144 14s 3d
Oxford Road	213	£0 12s 11.5d	£138 0s 6d
Old Street	97	£1 7s 10.0d	£134 19s 6d
Holbourn	169	£0 15s 7.6d	£132 1s 9d
Drury Lane	203	£0 12s 6.7d	£127 9s 9d
Goswell	101	£1 4s 9.6d	£125 4s 8d
Brick Lane	117	£1 0s 8.5d	£121 2s 9d
Smithfield	128	£0 18s 8.4d	£119 13s 9d
Butcher Row	181	£0 13s 2.1d	£119 4s 0d
Carey Street	163	£0 14s 0.2d	£114 5s 0d
Greys Inn Lane	128	£0 17s 6.3d	£112 3s 0d
Shoreditch Street	169	£0 13s 2.8d	£111 16s 1d
Golden Lane	103	£1 1s 6.3d	£110 17s 6d
White Chapel The Road	137	£0 15s 11.7d	£109 8s 4d
Low Holborne	144	£0 14s 11.2d	£107 11s 0d

areas had been important for the New River Company as far back as 1620. When similar plots are produced for revenue and user density, they reveal that Fleet Street and the areas north of St. Paul's were sites of more intense consumption even 150 years earlier.

The poorest streets, by contrast, were largely in the East End (table 6.6). The streets around Bishopsgate Street about the Old Artillery Ground, such as Artillery Lane and Skinner Street, had the lowest average rents. Most of the other East End streets with the lowest average rents are obscure. They were short, and their names have changed or disappeared with the years. Of the twenty poorest streets with at least ten customers on the New River's list, fifteen were in the east of London. Even so, the people on these streets were relatively well off to be able to afford water connections. These streets all had average yearly rental of more than twelve shillings, while the company had around eight hundred houses paying ten shillings. In effect, although the East End was clearly the poorest part of the city, the company's tenants with the lowest rates were not concentrated on any single street. In addition, there were a very few entities that got water for free, such as a poorhouse.[67] Despite the relative poverty of some parts of the East End, it was nevertheless, an area of sufficient demand that in 1805, the New River also built an iron main to meet growing needs.[68] The East End was becoming more important for New River after 1790, as it was for the Shadwell.

The effects of competition are also evident from the 1770 maps. The Chelsea Waterworks evidently had much of Westminster to itself as the New River could not supply that city except around Whitehall and the area to the north of St. James's Park. Although the New River could reach deep into the West End to the edges of Hyde Park, it took the most lucrative customers only in the areas close to Hyde Park, effectively leaving this part of the metropolis to the Chelsea. The smaller companies had little discernible impact in the West End, except perhaps the York Buildings Company around St. Martin's Lane, which was right next to where its pumps were located. The New River displayed a low point in service in its immediate vicinity. The LBWW by contrast, clearly dominated the New River in much of the zone where they overlapped. The New River barely registered customers in a region around the foot of London Bridge, stretching to St. Paul's in the west, the Tower in east, and up Bishopsgate to the north. This area of weakness corresponded almost exactly to Hearne's description of the extent of the LBWW's mains in 1745. Competition, therefore, was real in 1770 as it had been in 1710, when a report to the New River's directors complained of the

TABLE 6.6.
Streets with Lowest Average Rent (>10 tenants), New River Company

Street	Number of Tenants	Mean Rent	Total Rent
Cartes Street	10	£0 12s 0d	£6 0s 0d
White Street	10	£0 12s 4.8d	£6 4s 0d
Peter Street	19	£0 12s 9.5d	£12 3s 0d
Flower de Lace Street	10	£0 12s 9.6d	£6 8s 0d
New St. Giles	10	£0 13s 6d	£6 15s 0d
Old Short Road	27	£0 13s 6.2d	£18 5s 0d
Skinner Street	25	£0 13s 11.0d	£17 8s 0d
Queen Street	14	£0 14s 1.7d	£9 18s 0d
St. Dunstan Hill	11	£0 14s 2.2d	£7 16s 0d
Northampton Field	10	£0 14s 9.6d	£7 8s 0d
Susan Street Minos	12	£0 14s 10d	£8 18s 0d
Hopkins	10	£0 14s 10.2d	£7 8s 6d
Artillery Lane	31	£0 15s 3.9d	£23 15s 2d
Castle Street	19	£0 15s 6.9d	£14 16s 0d
Grosvenor Passage and Mews	10	£0 15s 7.2d	£7 16s 0d
Little Carter Lane	18	£0 15s 8.7d	£14 3s 0d
Cutter Street	10	£0 15s 9.6d	£7 18s 0d
Leadenhall Passage	17	£0 15s 11.8d	£13 11s 8d
Botolph Passage	10	£0 16s 0d	£8 0s 0d
Sclater Street	13	£0 16s 0.9d	£10 9s 0d

LBWW taking customers.[69] This pattern of usage also dates back to as early as 1620. Although the New River had by now extended its mains to the very east of the City, it mains did not go south toward the Thames in that zone, indicating that the LBWW could still hold its own against the New River.

Municipal Uses: Fire and Cleaning

Thus far, this chapter has dwelt on domestic and commercial consumption, but these were by no means the only uses made of water. There were also communal uses, most importantly firefighting and, to a lesser degree, cleansing the streets. Water was used for firefighting from the beginning of the water companies and had been put forward as a reason to establish them. When considering Peter Morris's project in 1582, the lord mayor wrote to the lord chancellor supporting it, stating that it should be promoted lest the help it offered "in cases of fire and infection . . . be lost."[70] The City's lease with the Broken Wharf Waterworks in 1604 allowed it to break its pipes to fight fires.[71] For their part, Hugh Myddelton's supporters encouraged the City to lend its aid to his project because its water would "be in a redines for danger of fires."[72] The New River's water was apparently put to use three times in 1615 to combat conflagrations.[73] When the New River was struggling unsuccessfully to get an act from Parliament in 1624, one of its supporters,

Sir Edward Coke, claimed that the company could help prevent "one great mischief that hangs over the city: *nimia potatio, frequens incendium*."[74] In the same year, the City's aldermen voted to grant Myddelton a chain of gold in recognition of "the great and extraordinarie benefitt and service this cittye receiveth by the water brought through the streets of the same by the travaile and industrye of Sr Hugh Middleton knight and Baronett especially att many great fires happened wthin this cittye, and chefely the last night at a verie terrible a fearefull fire, which might have greatly endangered this cittye, had that needefull water binn wanting."[75] The fire had destroyed three houses on Broad Street.[76]

During fires, the practice was to break open the pipes, but this soon led to considerable expense for the companies. In 1631, as it became increasingly clear that this would continue, Myddelton requested compensation from the City for the "great damages hee hath sustyned by the many and often breaches of the pipes vpon occasion of fires within this cittie."[77] He died before the matter was resolved, but in 1634 the City councillors resolved to pay his widow one thousand pounds for the damages sustained by the company, collected from fees to be levied on the wards. They also resolved that in the future firecocks or fireplugs should be placed along the mains at the discretion and expense of the wards. These firecocks were to be locked by keys that ward authorities would keep in their possession and use to open the mains for the "vse of the water in tyme of necessity vpon any accident of ffire."[78] Five years later, it was evident that not all the City's citizen's were willing to contribute for this shared cost, but the councillors ordered that those refusing to pay should be pursued in the courts.[79]

The first fifty years of water company operations effectively established a pattern for the relationship between the City and the companies for firefighting. The companies would provide water at no charge and would do all they could to ensure that it was flowing to where the fire broke out. For their part, the local wards paid for the installation of locked firecocks along the mains, which they had the keys to, and could open when needed. It was understood that these valves would not be opened except in the case of fires, at which point the same local ward official could open them and possibly attach a standpipe to the pipe. The water would then be used to fill buckets or, if the lay of the land was favorable, would be allowed to run freely along the street and diverted into the burning building by hastily constructed dams.[80]

Fire engines were also used beginning in the early seventeenth century. These were basins on wheels with pumps used to force water up a long pipe.

The engines were filled with buckets, and nearby water mains were a handy source.[81] This arrangement was not without its difficulties, nor was it always followed. The New River directors, for example, thought for some time before the 1660s that they bore too many expenses. Before the Great Fire, they sometimes petitioned the City for more money for "some recompence to be made for the great losse the Company hath susteynd by reason for quenching of fyers sithence the first bringing of the water."[82]

The Great Fire of 1666 changed firefighting in London in important ways. The London Building Act of 1667 set some standards to be used in new housing construction within the City, such as requiring all exterior walls to be of stone or brick and of a minimum thicknesses. Party walls were also required to be higher than roofs to slow the spread of fires.[83] Furthermore, fire insurance companies were founded. Mutual fire insurance groups had formed on the continent in the sixteenth century, developing into formal companies for the first time in Hamburg in 1676. In England, pre-1666 attempts to create insurance companies had failed. The Great Fire's destruction gave the subject renewed urgency, and the builder Nicholas Barbon founded the Fire Office with partners in 1680. Another followed in 1687, and both received letters patent that year. Others, such as the Hand-in-Hand, the Phoenix, and the Sun soon followed. These insurance companies hired people, originally Thames watermen, to fight fires that broke out in buildings they covered. These soon became formal brigades with their own distinctive livery.[84] Finally, improved fire engines were imported from Holland. Designed by Jan van der Heyden, they could suck water through a leather hose into the cistern, doing away with the need to fill it with buckets. The outflow also squirted through a leather hose under higher pressure rather than a long metal pipe, allowing the operator to stand farther from the fire. The addition of an air vessel in the 1710s forced the water out in a constant stream over longer distances rather than in weak bursts.[85]

In 1707, prompted by a rash of fires in London, Parliament passed an act mandating that all parishes within the London bills of mortality (an area covering the City, Westminster, Surrey, and many surrounding parishes) have fireplugs on mains and mark their locations clearly with signs on adjoining buildings. They also had to have fire engines and rewards were to be paid to the first "keepers" to arrive on the scene of a fire.[86] Although this helped in fighting fires, problems with this arrangement soon emerged in other ways. The New River directors were vexed by churchwardens who helped themselves to water through these plugs, or even gave the keys to friends. Some even removed the locks entirely so anyone could draw water from the pipes with their hands, and they wasted "vast quantityes whereby the owners of such water works are deprived of water for their customers." The company had to sue people to prevent the practice.[87]

To what degree any of this development made London's firefighting capacity more effective than that of other cities is not clear. By the eighteenth century almost the entirety of the city was crisscrossed with pipes, oftentimes with many provided by different companies in a single street. This state of affairs certainly represented a significant degree of water availability. There is anecdotal evidence in the form of contemporary observers praising London's firefighting capacity. In 1725 Daniel Defoe stated:

> No City in the World is so well furnished for the extinguishing Fires when they happen.
>
> 1. By the great Convenience of Water which being every where laid in the Streets in large Timber Pipes, as well from the Thames as the New River, those Pipes are furnished with a Fireplug, which the Parish Officers have the Key of, and when opened, let out not a Pipe but a River of Water into the Streets, so that making but a Dam in the Kennel, the whole Street is immediately under Water to supply the Engines.
>
> 2. By the great Number of admirable Engines, of which, almost, every Parish has One, and some Halls also, and some private Citizens have them of their own, so that no sooner does a Fire break out but the House is surrounded with Engines, and a Flood of Water poured upon it, till the Fire is, as it were not extinguished only, but drowned.
>
> 3. The several Ensurance Offices, of which I have spoken above, have each of them a certain Sett of Men, who they keep in constant Pay, and who they furnish with Tools proper for the Work, and to whom they give Jack Caps of Leather, able to keep them from Hurt if Brick or Timber, or any thing not of too great a Bulk, should fall upon them.[88]

Defoe further claimed that Londoners were careless about causing fires, probably owing to a sense of security on this score. Foreign visitors made similar comments, such as Johann von Archenholz in 1785: "[The water pipes] are extraordinarily useful for blazes, since as soon as the standpipes are set onto the opened pipes, they give water in quantity because of the constant flow of water therein."[89]

A less frequent and unsanctioned use of water was cleansing and watering the streets. Although it had been mentioned in some of the original water acts and patents, companies were resistant to letting their water flow freely on road surfaces simply to cleanse them. The interest in doing so was the idea this could prevent the plague, as argued by Francis Herring in 1625: "Let the pipes layd from the new River be often opened, to clense the channels of every streete in the Citie. Let the ditches towards the suburbs, especially towards Islington and Pick-hatch, Old-streete, and towards Shoreditch and White-chappell, be well clensed, and if it might be, the water of the new river to runne through them."[90] The streets were also watered during the summer to keep the dust down. This could be done by carts bearing water-filled barrels with holes through the streets.[91] At other times, the water companies opened their fireplugs and let the water run on the streets, such as in the summer of 1800, when the lord mayor asked the LBWW to freshen the streets in the mornings and evenings.[92] The water companies usually objected to the parishes taking the initiative and opening fireplugs for cleansing streets, insisting they should be opened only for fires.[93]

Conclusion

From the point of view of consumption, the London water industry took off after 1660, acquiring tens of thousands of customers to 1700. It was during this period when it grew most rapidly relative to its existing size, although not in terms of absolute numbers. This period coincided with the coming of internal peace following the Restoration and many other phenomena that, taken together, made for a heady time in the English economy. The Glorious Revolution, the financial revolution, and the age of improvement and projects were some of the changes and movements that occurred at this time. For the purposes of this chapter, it was the late seventeenth-century consumer revolution imported from Holland, however, that was the most important in sustaining the water industry's growth, especially when coupled with London's dramatic late seventeenth-century growth spurt. The industry in London had been firmly established by 1650, but it was not yet a mass-market phenomenon. The number of houses connected shot past ten thousand sometime before 1670 and reached around thirty thousand in 1710. What sustained this growth was the growing wealth of Londoners, as well as their desire to consume water in their houses. The contrast with Paris reveals that city size and technology alone were not sufficient to produce this result. The shift in consumer attitudes that saw the availability of water in the home as desirable and necessary was the foundation of the demand. Londoners could have continued to fetch water from public fountains, wells, or even standpipes on mains, but after 1650 they chose in huge numbers to have domestic water connections.

The new attitude formed during this consumer revolution proved enduring. The eighteenth century followed the pattern established after 1650. Demand for water was sustained and growing. The slowing of London's expansion between growth spurts allowed the water companies to consolidate, and market penetration rose relentlessly from the 1740s. Other building booms in the 1770s and 1800s slowed its rise, but this merely confirmed the presence of strong demand. By the 1820s, almost 85 percent of houses in London had piped water, the culmination of changes inaugurated by the consumer revolution 170 years earlier. The increasing market penetration of the London companies reflected that piped water was becoming ever cheaper in real terms compared to wages, and from being a privilege reserved for the wealthy in 1660, it had become available to a far greater proportion of the population in 1820. The wealthy West End dominated the geography of consumption for most of the eighteenth century, but consumption slowly extended to less wealthy segments of the population. The East End saw much more service from 1790 when the number of customers of the Shadwell Waterworks had increased rapidly, passing both the LBWW and the Chelsea Waterworks, a trend dramatically strengthened with the advent of the East London Waterworks after 1806. The New River also increased its supply there, finally adding a new iron main. There were, to be sure, vast numbers who got water from wells far into the nineteenth century. Another century and many epidemics would pass before full water availability came to London.

CHAPTER SEVEN

Purity, 1700–1810

Is it proper, that thousands of industrious people, who are obliged to be
in the streets, and fields, about their business, should be kept in continual
terror, because every lounging fellow must have a cur or two at his heels,
to throw into the New River, whenever he goes strolling into the fields?

A well-wisher to the human species, *Public Ledger*, August 20, 1760

[New River] waters may with safety and propriety be used, wherever
a pure soft water is requisite, for drinking or bathing; for washing or
bleaching; for dressing of food, animal and vegetable; in the ways of bak-
ing or boiling; for making malt and for brewing; for preparing medicines
by infusion, decoction, distillation, &c.

Charles Lucas, *An Essay on Waters*, 1756

The purity of London's water would become a subject of intense debate in
the nineteenth century when the growth of industries with noxious effluent,
such as the gas industry, and the intensification of water use, especially with
water closets, meant that both local wells and water bodies became increasingly polluted with organic and industrial waste. The various responses to
these problems at that time would include the public health movement and
the construction of large-scale water and sewerage infrastructure.[1] In the
context of London in particular, Edwin Chadwick's sanitary reform movement in the 1840s led to the creation of boards of health, followed later by
the construction of Joseph Bazalgette's great sewer network.[2] The issue of
water quality and means to maintain its purity were, however, also debated
in London before the transformations would put the issue into a new, more
pressing, perspective by the pollution that would produce the "Great Stink"
of 1858 when the Thames itself became an open sewer.[3] The question of the
purity of drinking water was of active interest, both to the water companies
and in public discourse, in the decades immediately preceding the debates
that flared in London over water quality beginning around 1810.[4]

Interest in London's water quality, particularly that supplied by the New
River Company, by far the city's largest water company, was intense in two

communities in particular. One was the corporate world of the New River Company, which was actively concerned with the purity of its water. It manifested this concern by employing many people to ensure that it served what it regarded as uncontaminated water. This interest on the part of the company demonstrates that active solicitude over water quality in water companies did not arise in the early nineteenth century with filtration but was operating earlier, albeit in a very different way that reflected a contemporary understanding of purity. The company was concerned less with filtration, and much more with preventing impurities that impinged on the taste and odor of the water. This meant especially leaves, weeds, and mud, but from the 1780s, likely influenced by changes in medical thinking, it became increasingly concerned with animal matter and then, beginning in 1800, with privies located close to the river.

That the broader public was also interested in water quality was indicated by a good deal of public commentary about the quality of water supplied by the New River and other water companies. In a series of sporadic outbreaks of sometimes vigorous complaining, anonymous writers voiced anxiety about New River water quality in letters published in newspapers. Their worries did not, however, entirely coincide with how the company expended its energies in keeping its water pure. While the company was mostly interested in preventing matter entering the water and thereby discoloring it and giving it a notable flavor, public debate was centered on people bathing in the river. Despite the thousands of cows that pastured next to the river, or even the privies that were occasionally sited nearby, it was people washing and swimming in the river that attracted sometimes bitter complaints and calls for the company and local authorities to act. In responding to these calls, local magistrates were known to pursue such offenders, sometimes even imprisoning them.

The New River Company's Efforts to Maintain Water Quality

During most of the eighteenth century, discussions about the quality and salubrity of water in medical and chemical texts were largely concerned with waters from baths, spas, and natural springs. This focus had its roots in the seventeenth century when, influenced by continental discourse and interest in spas, they once again became popular in England, accompanied by the publication of treatises discussing specific spa waters and their curative potencies. Holy wells and baths had fallen out of favor in England during the Reformation because of their frequent association with shrines.[5]

Before the nineteenth century, people's judgments about the quality of drinking water were based largely on evidence directly from the senses, rather than any kind of analysis. From antiquity, it was judged that if water smelled foul, tasted bad, or was visibly turbid, then it was of lower quality.[6] This aversion was at least in part prompted by the idea that rancid smell could cause illnesses, but this was by no means the only factor.[7] Physicians and others recommended in various ways that foul smelling or muddy water should be avoided.[8] Distilled water was generally regarded as the purest sort. However, because virtually all water was thought to be capable of putrefying or decaying matter, it was generally held that water in a state approaching this purity was hardly ever to be found. Water would inevitably cause putrefaction and fermentation in the materials it came into contact with, and even over time would purify itself as these processes would consume the matter mixed in with the water.[9] Some foreign matter, such as leaves or weeds, was easily separable from the water, being extrinsic to it. Other matter, such as minerals that caused hardness, were regarded as intrinsic to water and not easily separable from it, instead requiring chemical action.

Informed by these attitudes, the New River Company was concerned about the quality of the water it supplied and adopted various strategies, which evolved with time, to keep the perceived quality of its water high. Specifically—and in addition to its war on bathers—the company followed a two-step strategy to maintain quality. On the one hand, it tried to prevent the water's contamination particularly by vegetable and animal matter, while on the other, it allowed the water to sit in ponds so that impurities could settle out of it. From the 1780s, the company also began to show more care in preventing animal and other contamination from entering the river, and from 1800 in particular, it became much more actively worried about the presence of privies close to the river.

Details of how the company tried to avoid contamination of the river are sparse until the 1760s. What is clear is that even before this date, the company was fairly vigilant in regard to preventing contamination of the New River aqueduct itself. There were relatively few precedents for this sort of vigilance partly because the company had exclusive rights over the New River channel from its 1619 charter.[10] In other circumstances where common water use was an issue, such exclusive rights did not exist. Typically, infringement on common water rights would have to be dealt with through public nuisance suits in common law at quarterly assizes. In towns, public nuisances were sometimes administered through courts leet. These were public bodies of medieval origin that met occasionally and had responsibility for the presentment of nuisances. If the court leet noted a nuisance, it would be referred to a nuisance officer to prosecute, but no action could be guaranteed. Larger towns could have special petty courts, such as the court of conservancy in London that had responsibility for fisheries and hence could investigate and issue nuisance complaints. In addition to nuisance prosecutions, commissions concerned with river navigation improvement had existed since the late Middle Ages, originally in the form of commissions of sewers. These usually had limited authority, and their members were riparian landowners. Permanent river improvement or navigation commissions were more common from the seventeenth century.[11] In the London area, the Thames Navigation Commission was created permanently in 1751, following earlier temporary ones. River commissions could attempt to remove obstacles, repair banks, build locks, prevent water abstraction, and the like, although they could face serious opposition from landowners, many of whom were on the commissions themselves. Much of the New River Company's activity in cleaning the river, such as dredging mud or removing trees, was similar to what navigation commissions were doing to improve water flow, but with the difference that the company had exclusive access to the channel.

Evidence for the company's concern about water quality dates to the seventeenth century. The company's original charter included a section forbidding the casting of "anie unwholesome or uncleane thinge" into the river, as well as forbidding people to "washe nor clense anie Clothes wooll or other thinge" there.[12] This was evidently not sufficient, because in 1686 a royal proclamation was issued at the company's request and reissued in many subsequent years. It enjoined people not to "disturb, infect, abuse or corrupt the same river, by setting up of gates, overflowing of grounds, making of trenches or drains, filling of ponds, fishing, watering of cattle, keeping of geese or ducks thereupon, casting of carrion thereinto, or by doing or permitting any nuisance, annoyance, let, stoppage, or prejudice whatsoever to the stream."[13] In 1688, soon after this proclamation was issued, the directors became concerned that a Mr Newman was running three sinks into the river. They resolved to sue him unless he stopped doing so.[14] Besides responding to specific incidents such as this, the company began active efforts to prevent these abuses by employing walksmen to patrol the river constantly over its length. The number of these varied over its history, but during the eighteenth century it was usually between twelve and fourteen.[15] Besides repairing riverbanks and guarding against theft, the walksmen were employed to "prevent the throwing of filth, or infectious matter" into the river.[16] In addition, they also removed weeds from the river and cut back vegetation along its banks.[17] The weeds were considered a nuisance because in addition to the flavor they gave the water, they slowed the current and could clog grates and pipes.[18] Walksmen deemed negligent in cutting weeds could be dismissed.[19] As for the trees, their roots were problematic because they could undermine the river's bank, causing leaks, and slow the river's flow.[20] Moreover, their leaves falling into the river could also create blockages and flavor the water, and so the walksmen also cut back any branches overhanging the river.[21] As well as debris, the company regularly removed mud from the river and its reservoirs, largely to prevent them from silting up.[22]

Medical writers and the New River's consumers expressed a range of opinions about the quality of New River before the 1750s. Many were positive, such as Thomas Tryon, an early proponent of vegetarianism, who in 1683 wrote that "the New-River that supplies London, is some of the best water in England (except Thames water) it being a cut or made river that runs of the surface of the earth."[23] In 1701 James Harvey preferred New River water over Thames water, which he claimed had been vitiated by dyers: "No water is fit for bread, save the New River."[24] John Hancocke, a cleric, went even further than these earlier statements. Hancocke was a proponent of treating illnesses using pure water, and the New River water qualified for this use, although not as fully as spring water. In his 1722 book *Febrifugum Magnum*, he claimed that, although many waters could be used in treating fever, he preferred pump water over the New River's, which was "often not so clear and sweet."[25]

Not all opinion was positive, however. In a book from 1652, Nicholas Culpeper, an influential physician engaged in public education in health matters, described New River water as "muddy."[26] Some brewers noted its muddiness as well. William Ellis, for example, stated in 1734 that although "the New-River [water] . . . is the best sort that London affords for brewing," it still contained "vasy, muddy sediments" which "subsided at the bottom" of the cisterns used to collect the water at breweries.[27] Henry Baker, a teacher and natural philosopher who made microscopic studies on freshwater polyps in the 1740s, was also negative about the mud in the water, although he thought the water could easily be purified: "[It] may be rendered much better . . . if it be let stand for a day or two, till the foulness subsides, and it becomes perfectly fine and limpid."[28]

From 1769, details about sources of contamination of the New River and how the company dealt with them become clearer, and there was an evolution in what the company was concerned with. Originally, animal matter was not regarded as much of a problem. Because the New River ran for tens of miles through farmland, exposure to agricultural contamination was always possible. Before the 1780s, there is little evidence that the company tried to keep cattle very far from the river. There were thousands of cows in the area of Lea Valley since it was an important agricultural zone supplying London. On the contrary, "watering places" for cattle along the river are mentioned in the records. In a case from 1744, when cattle were becoming mired in the river, the court allowed the farmers to fill in the bank and create a watering place.[29] The concern was that the cows were being injured, not that they were contaminating the river, despite the spread of cattle plague in Europe at that time.[30] Similarly in 1769, the company requested that a farmer adjoining the channel near Islington repair his "cowlayers" because the barriers between them and the river had broken down. When he had not done so after a few months' time, the court of directors ordered that the river's banks be restored at its expense.[31] The problem was once again the state of the banks, with no mention of cattle contaminating the water. Likewise, in 1780 the company wrote to William Marshal, asking him to move his hog sty next to the church in Hornsey because it was too close to the river. When Marshal visited the court to ask for a reprieve, it was refused, as "the company would not suffer any nuisance or encroachment on the banks of the river."[32] In general, it was clear that the company was willing to allow a connection between cow ponds and the river so long as the farmer paid for his water.[33]

Despite this earlier insouciance about livestock, there was a slow change in the company's attitude toward animal pollution as revealed by its actions after 1780. What caused this shift is not evident. One possibility is that medical writers wrote more frequently about water-related illnesses, although this was sporadic and not enough research has been done on this subject to make links clear. One influential writer of the time was the physician John Pringle, who began to investigate putrid effluvia as possible causes of illness. He thought putrefying animal matter was a disease-causing agent, a point he argued in his 1752 work on dysentery. He claimed that the illness was caused by poor ventilation and that contact with putrefying animal matter could lead to diseases.[34] Pringelian interest in the environmental context of disease was mirrored by a few physicians who discussed New River water. One was William Heberden, a wealthy and influential London doctor. In 1767 Heberden read a paper to the Royal Society on London's water supply and used strong language to describe the New River Company's water. He specifically identified it as a source of ailments, particularly related to London's frightful rate of child mortality: "There is an inconvenience attending the use of Thames and New river water, that they often are very muddy, or taste very strongly of the weeds and leaves. The latter fault is not easily remedied; but they would soon be freed from their muddiness, if kept some time in an earthen jar. If the water given to very young children were all of this kind it might perhaps prevent some of their bowel disorders and so contribute a little to lessen that amazing mortality among the children which are attempted to be brought up in London." While the observation about the New River water's muddiness was by this time commonplace, Heberden's linking of it to ailments was not.[35]

A further medical opinion on New River water came in 1773 from Charles White, a Manchester-based surgeon who played an important role in the medicalization of childbirth in the eighteenth century. He wrote about putrefying animal and vegetable matter in New River water in the context of childbirth.[36] A few years after Heberden's paper, he identified the extensive use of New River water as an important cause of diseases in childbirth and of other ailments: "Is it not one cause of the frequency and fatality of the puerperal, jail, hospital, and other putrid fevers, in London, that so many of the inhabitants drink, and use for most culinary purposes, the New River water, which is frequently replete with putrid vegetable and animal substances?"[37]

John Fothergill, a London physician and prolific author known for his concern for the health of the poor, wrote a letter in 1780 to the New River Company itself expressing his thoughts about water quality. Like the doctors who had published about the New River's water in the preceding years, Fothergill identified putrefying matter as something that made the water less wholesome and further suggested that the sun helped purify the water.

Returning lately from the North and passing through Ware I was struck with observing the quantity of leaves falling into the New River from the trees growing along its banks at the same time reflecting that in the whole of its course to Islington the quantity must be such as could not sail by rotting in the current to render the water less pleasant and less wholesome. Whether this circumstance has occurred to you I know not but I am very certain it must have a considerable effect on the water and it appears from late experiments that the sun has much influence in meliorating the water. For both these reasons I imagine you will think it expedient to give particular instructions to those who have the immediate care of the river committed to them to take care to have as many of the trees removed from the verge of the river as possibly can be done without creating too much opposition to prevent every where any new ones from being planted so near the river as to injure it either by their shade or the falling leaves.[38]

The company, specifically citing this letter, tried harder to prevent leaves from falling into the river and asked at least one neighbor on the river whose trees were dropping leaves whether he could pay to have them cut down.[39]

To what extent this letter and the observations of Heberden and others had a direct influence is not evident. It is clear, however, that the company was more concerned with the dumping of waste in the 1780s than it had been earlier, as least as reflected in its minutes. In 1782, for example, the company wrote to Mr. Bell, asking him to take care of the "great nuisance" he was causing through the brick drain from his house near Cannonbury House that discharged "black and filthy water" into the river.[40] A little later, Bell stated to the company that it could do whatever it wanted at its expense, and so it ordered that the bank be rebuilt to cut off the inflow.[41] Similarly in 1787, the company found two drains from a house in Enfield running into the river, which it ordered blocked.[42]

Clearer examples of the company's concern about animal contamination came in the 1790s, although even at this time, the company was not vigorous in protecting from all contamination. In a case in 1791, the court moved against a neighbor because "the dung from a stable . . . is laid on the wharf of the New River . . . which occasionally drains and is blown by the wind into the River."[43] The initiative came not, however, from the company, but from people complaining in letters. The court asked the stable owner to cease putting the dung on the company's wharf. In another case from 1799, the Baron d'Aguilar, known to be miserly and eccentric, left a large pile of dung and filth on the river's bank. It was "very offensive, and much complained about."[44] The baron was not sympathetic to the company's complaints and did nothing about it, forcing the directors to have a "banck of clay" built to contain the filth.[45] In another episode from 1799, the court of directors wrote to Mr. Potter of Wood Green that "you make a custom of suspending or laying horse flesh in the New River with which you feed your dogs." The company threatened him with prosecution.[46] In another episode, "a quantity of butcher's meat" was found in the river in Enfield. The company offered a reward of five guineas to whoever gave information leading to a conviction of the offender.[47]

Connections between the river and cow ponds, however, continued to be tolerated during the late eighteenth and early nineteenth centuries. In 1793, for example, a farmer requested a supply of water from the river for his cattle, and the company agreed, ordering that a trough be installed linked to the river to supply water for this purpose.[48] Other cases of cows in close proximity to (and sometimes drinking from) the river were recorded into the nineteenth century.[49]

Another shift is evident in the early nineteenth century as the company began to show more unease about privies placed close to the river and human waste more generally. No reference to concerns about privies can be found in the company's minute books before 1800, but from that time the issue emerged on a regular basis. In a survey of the river that the company's directors made in 1802, several were discovered. In addition to some near the river in Enfield, there was one belonging to the company itself near Ware, which was ordered moved so as not to be "a bad example to other persons."[50] One of the Enfield privies was five feet (1.5 m) from the river, and the company rebuilt it farther away at its own expense.[51] In 1807, a neighbor at Dead Man's Hill had cut a ditch from the privy in his garden to the river, which the company demanded be closed.[52] In one of the most egregious cases, in 1809 John Cooper, an inhabitant of the village of Ware, became suspicious of the actions of some nightmen and hired a watchmen to follow them secretly after dark. He discovered that these nightmen collected waste from a privy and disposed of it in the New River. When the company was informed, it tried to track down the nightmen to prosecute them, but they had disappeared.[53] In another case from 1809, it was found that a privy next to a cottage in Enfield was overflowing into the river and that the cottage's inhabitant was in the habit of throwing dirty water into the river. This led to a heated exchange of letters before it was solved.[54]

Bathing in the New River

If the New River Company became increasingly worried about different sorts of animal and vegetable matter in the river in the second half on the eighteenth century, a broader segment of the population also manifested concerns about the contamination of its water that might result from people bathing in the river. . The existence of an intense discussion by this time signaled a change in the nature of how the public in London was willing to approach the question of the purity of its water supply. Although the New River Company had been in existence since the early seventeenth century, and supplying thousands of houses by the late seventeenth century, only from the 1750s onward is it possible to discern a consistent discussion in the public sphere about the purity of its water. While the nineteenth century witnessed the rise of public health in the context of large-scale concerns about water supply, the stage for this had been set by the rise of public debates during the late eighteenth century.

Why people bathing and washing during the warm summer months (and usually on Sundays) attracted far more attention from the public than all other kinds of contamination combined is not immediately evident. Mary Douglas, in her foundational work on the anthropology of pollution, argued that societies identify some things as "dirt"—forms of pollution—because they are regarded as matter out of place within a societal framework. For Douglas, pollution was not only about the material aspect of pollution but also its place within a societal system that regarded the offending matter (or behavior) as contravening that societal order. For the anonymous complainers who were so aghast at the filth of the New River bathers, the perception of bathing itself as a cleansing process had strengthened in the latter part of the eighteenth century, making bathing in London's potable water supply more of a "filthy" act, becoming a threat to order. Other forms of immorality presented by the bathers, notably nudity, the profanation of the Sabbath, swearing, and harassing of women passing through the area besides other crimes, all combined to confirm the moral depravity of the people involved. The physical contamination of the river was inextricably linked to the moral sort.[55]

How had bathing itself changed in the second half of the eighteenth century? It was a time when frequent, even daily bathing was becoming more common among the upper classes throughout Europe. While spa bathing had been popular in the late Middle Ages, and continued to be so to the eighteenth century, cold bathing at spas had not. In England, William Floyer had begun a revival around the turn of the eighteenth century, and this activity continued to grow in popularity among the wealthy elites as the century progressed.[56] Regular bathing at home enjoyed a similar surge in popularity. Few people bathed in the seventeenth century, and into the 1740s it was still uncommon to bath regularly. By the 1770s, however, this was no longer the case, and finally in the 1790s it was fashionable to take a daily bath. The trend was partly driven by Pringle's insistence on cleanliness to preserve health. Both spa and home bathing activities were, however, confined to the upper and middle classes. Bathing was then becoming more closely associated with health and cleanliness, strongly associating the act of entering water with leaving behind dirt and uncleanliness in way that had not been the case when regular bathing had not been deemed important.[57] River bathing had never diminished in popularity from the Middle Ages, but it was a form of recreation, and not perceived as a hygienic act. As the wealthier segments of London society changed their own bathing habits and more clearly associated it with health and cleanliness, they also changed how they understood the health and cleanliness of river bathing. The dirt they now perceived to be leaving in their baths was being left in their drinking water by the New River bathers.

There were also other elements to this criticism that help explain why it became so intense in the late eighteenth century. One of these was the class character of the dispute. Many of the letters remarked on the low class of the bathers, while their few defenders relied on the dearth of opportunities for bathing among poorer people as a reason to excuse the behavior. The dispute can then be situated within a changing broader context of class and propriety. Norbert Elias argued many years ago in *The Civilizing Process* that people in the West were increasingly under greater pressure to conform to codes of politeness and etiquette through the increasing interdependence and depth of social relations, motivated by the expansion of commerce and the authority of royal courts.[58] In regard to nudity in particular, Elias argued that it was problematic in the Middle Ages only when occurring between social strata, but with increasing democratization it was becoming universally so by the late eighteenth century.[59] This trend would become more evident in England in the early nineteenth century as the regulation of decency drifted from informal mechanisms to ones defined by law and enforced by officials. Exposure had been actionable under common law since at least the seventeenth century, but nude river bathing in particular was first banned by statutory laws beginning with the Thames Police Act of 1814. The act's proponents in Parliament partly justified it by "the want of deference in the lower classes towards the higher, which had increased so much in later years."[60] The Vagrancy Act of 1822 then banned indecent exposure generally, but despite the growing moral outrage motivating the law, evidence suggests that the problem was not worse in the 1820s than it had been in the 1780s. It was a rise in the standards of respectability that created the pressure for the new laws. Enforcement was provided by officials such as police, in addition to volunteers who hectored lower-class youths bathing in Hyde Park and other areas around the city.[61]

That bathing occurred before 1760 is evident from early company records.

In 1614, a year after the river came into operation, the company was paying workers "to kepe out swymers in Whitsonne holidays."[62] When Charles II issued a royal proclamation in 1685 against contaminating the river, there was no mention of bathing, but a new one from 1689 differs from the first in only one regard, that people "washing themselves therein" was now forbidden.[63] In 1722 the company's secretary wrote to the headmaster of a boarding school near the river, asking that he take care to prevent his boarders from washing in the river, and reminding him that some people had recently been sent to prison on this account.[64] The first evidence of public complaints, however, came from 1755 when the company itself, in response to a complaint it had received, printed a notice stating that it would prosecute anyone found washing in the river, as this caused "great annoyance to people passing that way, and . . . prejudice to the water."[65] Little more was heard about the issue until the period from 1768 to 1770, when a rash of complaints appeared in London papers. The first came from "a lover of decency" and was printed in the *Gazetteer and New Daily Advertiser*. The New River water was not fit for use, the letter claimed, because of the "nasty filthy fellows bathing" in the river. Ladies used to be able to walk in the area but were now prevented by the "indecency and impudence of these people." The letter went on to state that "no one knows what diseases these persons may have, and as the river does not purge itself by a flux and reflux of the tide, the water is certainly contaminated by this nastiness." Watchmen and prosecutions of offenders would remedy the problem.[66] The sentiment was echoed by another letter observing that this behavior "not only offends the eye, but renders the water we are to drink extremely filthy and disgusting."[67]

Letters in the following year pursued the conflation of moral and physical pollution. One observed that "two hundred vagabonds" were bathing in river, as well as gaming, "cursing and swearing in a most profane manner." The corruption of youth was part of their crime: the apprentices and tradesmen's sons among them would "in all probability . . . be ruined by that assembly of thieves." The peace officers should take care to prevent this scandal.[68] Another letter suggested that the company itself should hire "two or three stout men" to throw the bathers' clothes into the river to stop the problem.[69]

The bathers were not without their defenders. A riposte to these complaints argued that bathing was healthy and refreshing, preserving the lives of thousands who engaged in it. Those who bath in the New River were not sufficiently wealthy as to afford subscriptions to private baths and would be deprived entirely if prevented from doing so. The true indecency was not in the bathing but with the women who walked nearby. Nor was the corruption of the water a problem because the river flowed for long enough afterward to purify itself. The complaining should cease, the letter finished, until a cheap public bath was provided for those who could not afford the private ones.[70]

Letters also brought pressure to bear on the company more directly. In a letter in the *London Evening Post* from September 1766, someone complained of the indecency that took place on Sundays, that of fellows "washing their dirty hides and polluting the stream." Although many people have complained publicly, the writer noted, the company had not fixed this problem. The letter writer suggested that the company should have its "river keepers" who removed weeds patrol on Sundays. This was urgent because, unlike the Thames, which purified itself by the tide carrying its filth out, the New River was not purged in this way with its slow current and stagnant reservoir. The letter finished by threatening to switch to another water company unless the company were firmer in "preserving decency on the sabbath."[71] This final letter of 1766 came too late in season to have an effect, but when warmer temperatures brought a return of the bathers in 1767, there was action. Constables began apprehending people for swimming in the river, and some were sent to Bridewell Prison, with others given fines.[72] Although this silenced the critics for 1767, the bathers and the complainers returned the following three years, with much of the same tone of mixing moral and physical pollution.[73]

The bitter complaining continued every summer over the next years, including one incident where the New River Company's engineer was assaulted and another when two women were accosted by bathers.[74] The rhetoric reached a crescendo in 1778–79, when calls to action became ever more strident.[75] The company proprietors were denounced, for, although they were an "opulent body of men," they allowed their water to be "fouled by the indecent custom of filthy fellows, and filthy dogs, washing in it on Sundays." If a sense of duty was not enough to move the proprietors, then "your interest should prompt you to attend to this matter" as the water "is so contaminated, and . . . has not time to cleanse and purge before it is used."[76] Other letter writers proposed more extreme solutions, such as using press gangs to force bathers into the Royal Navy,[77] or throwing pieces of broken glass into the river to cut the bathers.[78] The constables took up the first suggestion and in July 1778 impressed many bathers and sent them to the New Prison.[79]

As with the earlier episode, some of the letter writers defended the bathers, and the debate circulated around the issue of class privilege. One letter expressed the hope that "the proprietors of the New River water, will have more sense, as well as generosity, than to debar those poor men and boys . . . the liberty of washing in the New River, provided they observe decency." The letter pointed out that threats to stop using the New River were entirely hollow, as all other possible supplies of water "has much more filth in it." New River water cleared, the letter stated, after it was allowed to sit in a tub, and was perfectly fine for drinking. The complainers were only seeking to "hinder the poor of every enjoyment," and they should encourage bathing because of its positive health benefits. The poor could not afford baths and should be allowed this comfort at least.[80] In a further letter from the same author, he suggested that the complainers were the ones committing an indecency by taking their wives to the area where the bathers were.[81] The letters continued to pour in through 1779, with many calls to press the offenders into service.[82] Finally, the New River Company and the Finsbury constables began an energetic campaign to prevent the bathing, regularly collecting people caught in the act and having them condemned to prison. Some were also pressed into naval service. The papers contained many reports of bathers taken up like this in July and August 1779, in groups as large as twenty-five.[83] Perhaps because of these mass arrests, fines, imprisonment, and forced naval service, complaints almost ceased the following year,[84] although the company still put up signs threatening would-be bathers.[85]

The next burst of complaining about the purity of the New River's water ran from 1782 to 1786 and, like the earlier episode, focused on bathing and the connection between immorality and pollution. The first letter came in 1782 in the *Public Advertiser*, stating that "there is not a more offensive nuisance practised than the bathing in the New River on Sundays," and that genteel women were insulted by it.[86] The following summer saw the complaints culminate again. The first letter denounced the magistrates and the company for doing nothing to prevent the "indecencies" and "atrocious acts of villainy" and suggested it was because "they have neither wives nor daughters."[87] A sarcastic letter then suggested that the company, "in order to do every thing in their power to sweeten the water at this warm season of the year, permit men and boys continually to wash themselves in it, thereby infusing proper purgatives to render the liquid seasonably wholesome."[88] Further letters pressured the local constables and the company "to suppress that indecent practice of bathing in the water, which should come to the inhabitants of this metropolis pure, but which has been polluted, by an act which was at the same time a breach of the Sabbath, and an offence to decency."[89] A group of forty-two residents even wrote the company condemning the bathing, claiming that it made the water "thick and unclean" and that it caused "prejudice to their health and constitutions."[90] Finally, after all these letters, newspapers reported that the high constable of Finsbury and several peace officers took five bathers to Bridewell Prison.[91] The company also built a brick wall around the reservoir serving the residents who wrote the company directly.[92]

The letter writers were not, however, satisfied with this report. They claimed that the bathers were still there, despite what magistrates said, and suggested starting a subscription to hire people to patrol the river.[93] They also claimed that the New River Company was not as careful as it had previously been: "The New River Company I remember formerly used to be much more strict than they are now in preventing such enormities."[94] This cycle of condemnation of the bathers, hectoring of public authorities,[95] and finally action was repeated during the following three summers until constables patrolled the river at the most frequented spots leading to indictments.[96] The issue arose again sporadically over the following years to 1810, although not with the same intensity as the earlier bouts.[97] The usual remedies of prosecutions were applied, including with the help of a subscription raised from among the neighbors to hire watchmen to patrol the river.[98]

Conclusion

Although the salubrity of the water supply did not receive the centralized and bureaucratized attention that would come in the nineteenth century, nevertheless the history of the New River reveals that historical actors— individuals, the company itself, local authorities—could be moved to sustained action by health concerns in the eighteenth century. For its part, the company tried to mitigate whatever most affected the taste, smell, and aspect of its water. This meant removing mud from the river, cutting weeds, and pruning back trees to ward off leaves falling into the channel. Doing so required the employment of around a dozen walksmen to patrol parts of the river daily to do these tasks. The company was able to maintain this degree of vigilance because it had exclusive rights to its channel through its charter of incorporation, giving it a relatively free hand as compared to the riparian landowner commissions with authority over England's rivers. Although these tactics were in vigor throughout the century, a slow evolution was discernible from the 1780s in the company's concerns. Perhaps in reaction to the sporadic comments from medical authors on the link between animal and vegetable matter and illness, the company showed more care about these materials entering the river and from 1800 was much more concerned about privies, a subject that had not appeared in its minutes before that date. If London's mortality rate declined in the second half of the eighteenth century, it may be that this slow change had some contribution. Whatever its ultimate effectiveness, systematic and organized care for water quality existed in the eighteenth century.

Individual customers were also concerned about the quality of the New River's water. Occasional references in various texts before the 1750s remarked on the water's muddy quality. But from that point forward, the pollution of the river by people bathing in it repeatedly generated storms of complaint and controversy, playing out in London's papers. This source of contamination, more than any other that could have attracted attention, such as the thousands of cows pasturing along the river's shores, became a lightning rod for disaffection. That this was so reveals the interaction of a number of trends. From almost nonexistence in 1750, daily bathing (and going to spas) was becoming much more popular among the middle and upper classes as they more closely associated the practice with healthy living. The practice of bathing in the New River, which was not new in the late eighteenth century, then became a health-associated act. Bathing in the river left filth behind, whereas earlier it was simply swimming. Furthermore, the complaining letters reveal that their authors conflated the physical contamination of the river with the moral degradation of public nudity and other behaviors. It was this trespassing of moral boundaries that set off alarms in regard to "dirt," in Mary Douglas's meaning, being left in the river. Finally, there was a class element to the dispute as public nudity was increasingly being frowned upon, particularly by the upper classes. This would soon lead to statutory criminalization of the behavior in the early years of the nineteenth century. In the New River's case, the complainers were wealthier, while the bathers were lower class. The few people who defended the practice called for the provision of public baths for the poor, something that would come only in the nineteenth century.[99]

Historians have largely regarded the public health movement as a nineteenth-century phenomenon, particularly in regard the question of water purity. This history of the New River, however, suggests that concern about and action to preserve the purity of water and at least indirectly the well-being of its consumers was present in the eighteenth century as well. In contrast to the nineteenth century, when governmental bodies and laws were central to the public health movement, most of the initiative and action from the early eighteenth century lay within the New River water company. Only with the emergence of discussion in newspapers from the 1760s did the initiative shift from the company to a broader public, suggesting that the rise of the public sphere was important for the emergence of public discussions about the quality of water in London. The focus of these public discussions shifted in the early years of the nineteenth century as water quality debates flared again in regard to the purity of Thames water and the reliability of supply. As with the bathing debate, however, these new battles were fought in the public sphere. Finally, expert medical opinion was increasingly important in the late eighteenth century. Medical practitioners were more interested in the quality of water and in the water-related mechanism of disease transmission. The response of the New River to Fothergill's letter, as well as its changing attitude to animal and vegetable contamination and finally to privies by the end of century, points to sensitivity to medical opinion.

CHAPTER EIGHT

Epilogue and Legacy, 1800–1820

To be sanctioned by act of Parliament: A national Light and Heat Company for providing our streets and houses with hydrocarbonic gas-lights,
on similar principles as they are now supplied with water.
        Frederick Winsor, promoting the first London gas company, 1807

The New River Company is in a "critical and alarming crisis."
                                        Internal company report, 1814

The turn of the nineteenth century brought with it profound changes for
the London water industry. Some of these, such as the creation of large new
companies that posed serious competitive threats to the existing ones, were
truly novel phenomena. Other changes had been building for some time.
Among these longer-term changes were pumping by steam engines and the
use of cast iron pipes, joints, and valves. Both had become sufficiently reliable and familiar that investors could be confident of creating new companies
depending only on them. The confluence of these factors after 1800 meant
that within twenty years the water industry in London had undergone its
deepest changes since the rapid expansion of the late seventeenth century.
Within two decades, wood pipes were abandoned in favor of iron, and steam
engines became central for water supply. These two changes in turn meant
that high pressure and constant supply had been introduced, albeit nowhere
close to universally. The New River, while still the largest company, had lost
its overwhelming position, its market share dipping below half of the market
around 1812, a position it had held for about two hundred years. It continued
dropping to under 40 percent by 1820. The geography of supply was recon-
figured, with companies that had not existed before 1805 having large areas
of supply for their exclusive provision.

Transformations in London to 1820

The advent of war in 1792 following the French Revolution was one of the
triggers for the post-1800 changes. It stretched one element of the water sup-

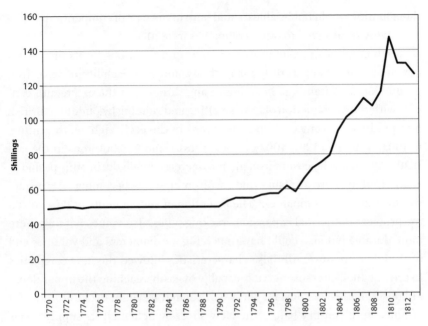

Figure 8.1. Price of wood in shillings per load paid by the New River Company, 1770–1812. Data taken from sales listed in MB1–9

ply model to the breaking point: the use of wooden pipes. Water companies had been using wood for more than two hundred years, and their experiments with iron had been limited to a few mains. As described in chapter 4, the price of timber for pipes had effectively held constant throughout the eighteenth century, and so there was no price pressure to change. The tumult of the wars caused wood scarcity and drove prices upward from around 50 shillings per load before the war, to 58 shillings in 1795, 67 in 1800, 96 in 1804, and 150 in 1810 (fig. 8.1). Given that wood was the water companies' largest expense, this effectively broke their business model. The London Bridge Waterworks, for example, observed the situation with alarm in 1807 and decided to move to iron pipes because elm was becoming too expensive and hard to find.[1] When investors were forming new companies, such as the West Middlesex Waterworks in 1807, they briefly considered using wood pipes, but they too quickly settled on iron.[2] The New River held out longer. As it came under pressure from both competition and wood prices, it finally decided to abandon wood in 1814. Over the next five years, it replaced most of its wood network with iron, with the exception of areas the directors thought they would not be able to hold onto in the face of competition. These areas were transferred to other companies in 1820.

The high price of wood was one of the central reasons for its final abandonment, but others were the availability and practicability of iron. Iron production in Britain had been increasing throughout the eighteenth century with the introduction of coke smelting and refining by puddling. British iron production went from to 20,000 tons in the early eighteenth century, to 100,000 tons in 1791, 400,000 tons in 1810, and 700,000 tons in the late 1820s. Although the cost of iron did not decrease much during this period, it was definitely more available, and cast iron pipe was becoming very attractive for the water companies.[3] The invention of sliding joints in 1784 solved the problem of thermal expansion cracking pipes.[4] Iron pipes not only were more durable but also could have much higher diameters and volumes and bear higher pressure. All these characteristics allowed the new companies to provide high-pressure service, capable of easily reaching the upper floors of houses, and even constant service. Although it would take until the end of nineteenth century before the New River switched all its supply zones to constant flow, the new companies began offering it right away. The slow transition to constant supply was due to the significant technical problems in adjusting networks to constant high-pressure supply, both for the company and for consumers, who had to adjust all their interior plumbing. For example, the ballcocks that had served as flow regulators for more than one hundred years could not work with high pressure, and new more expensive valves were needed.[5] In many cases, they were happy to keep their intermittent supply. In addition, water engineers thought the volume of water was simply not available.[6]

The ongoing development of steam engines provided the possibility of high-pressure supply. The New River could continue to rely on its gravity-fed water supply, but the steam engine had removed its lock on cheap water and high volumes. In the east of the city, the Shadwell Waterworks was already expanding rapidly before 1800. Between 1800 and 1820, the proportion of London water company customers getting water provided by steam engines rose from 35 percent in 1800 to 60 percent in 1820. Historians had pointed out that as iconic as the steam engine was for the industrial revolution, its application in industry remained relatively modest compared to waterpower in particular.[7] This was not true in the London water industry. Steam engines and iron, another central facet of the industrial revolution, were keystones of its early nineteenth-century transformation.

The new business landscape in the industry was enabled because of broader changes in the business environment. Joint-stock corporations were enjoying a renewed bout of popularity. The requirement to seek a parliamentary act for incorporation after 1720 ushered in a long period when the unincorporated company was the preferred joint-stock model. Matters began to shift with the advent of the canal boom from the late 1760s. There were a couple of waves when many new joint-stock canal corporations were created by Parliament. These typically were promoted and funded by local investors who needed rights of expropriation or compulsory access to land to construct their canals, which were granted in the acts. In addition, their high capital costs, as with water companies, pushed the promoters toward the joint-stock form. By the early nineteenth century, canal corporations were no longer a novelty, and their infrastructure crisscrossed large swathes of the country, driving down transportation costs and generating good profits. They had effectively shown the potential of joint-stock corporations for infrastructure construction in the new era. A new joint-stock boom followed from 1805 to 1811. In the peak year of 1807, fully forty-two new companies were created. Not since the South Sea Bubble of 1720 did so many people seek to buy shares in corporations. The boom was present in wide sectors of the economy and not limited to canals, as had been the previous booms of the 1760s and the 1790s. Gas companies, breweries, vinegar manufacturers, dock companies, and of course water companies were all featured. The unincorporated joint-stock company, never properly recognized in statutory law, was fading as a business form.[8] The broad base of the new joint-stock boom was also evident with the new water companies. Whereas the New River in 1814 still had a mere 120 shareholders holding the 72 original shares between them, the new companies had by that point sold 12,700 shares, many to people with small holdings.[9]

The political interest in creating new companies was based on the rapid expansion of London and the sense that the existing water companies were not keeping up. The city's population grew from 960,000 in 1801 to 1,140,000 in 1811 and then to 1,380,000 in 1821. Many new areas such as in the east or the northwest of the city were not getting supply.[10] The acts creating the companies explained in their preambles that there was need for more water in some areas, such as Hammersmith. Far from stating that the purpose of new companies was to create competition, the acts could even contain provision to limit it. The West Middlesex Waterworks, for example, was restricted from entering the areas supplied by existing companies. The new company could not enter Westminster and Chelsea because, if it did so, it would have to pay the company already operating there, the Chelsea Waterworks, ten pounds for each house it took on. The Chelsea Waterworks had successfully lobbied Parliament for this provision while the bill was being considered. The East London Waterworks was not restricted in this way, but its act did contain a clause stating that nothing in the act should hinder the New River's rights. It was not more precise than this, and similar clauses had been used even in the seventeenth century. In practice, none of this limited competition.[11] A further motivation for creating new companies was mentioned in a New River's report written in 1814 on the effects of competition. The authors of the report argued that because the dividends of the New River had never been made public, and share sales at auction were reported at tens of thousands of pounds, people assumed that the company was making exorbitant profits.[12] In fact, the dividend returns on the shares for those who bought them were around 3–4 percent in the late eighteenth century, but for those who had inherited them, they were a comfortable source of income that required no investment of external capital, at least until the competitors appeared.

Soon after 1800, a series of new water companies was created. The first was the South London Waterworks in 1805, which was renamed the Vauxhall Waterworks in 1834 before merging with the Southwark Company in 1845. The West Middlesex Waterworks followed in 1806. Its charter allowed it to supply areas in the West End, which it did from pumps in Hammersmith. The East London Waterworks was formed in 1807, and it purchased the water assets of the Shadwell Waterworks and West Ham Waterworks in 1808 from the London Docks, which had taken over both the previous year. Both of the older companies had lost supply areas when the docks had been built. The sale of the water assets gave the East London Waterworks a significant step up in terms of size, and it soon pushed into the New River's supply zone in the east of the City. It pumped water from the River Lea. The last major company within the metropolis was the Grand Junction Waterworks. It was originally created by the owners of the Grand Junction Canal to sell the supposedly abundant water flowing into London through the canal. These aspirations were unrealistic, and the waterworks company was split off from the canal in 1810 to pump from the Thames instead. It was incorporated the following year and began operations in 1812. It was the first to use high pressure and constant supply. Its supply zone was also the West End, which, like in the late seventeenth century, became an area with many overlapping supply zones. The new companies were soon competing fiercely with the old ones and with one another. They frequently offered water service with no connection fee, thereby allowing customers to switch as frequently as they liked when offered lower prices. Agents from the new companies went door to door, canvassing for new customers. More aggressive tactics included breaking opposing companies' pipes in the streets.[13]

Losses soon built up at the old companies, finally forcing them to modernize their networks by rebuilding them in iron over the course of a few years. They also sought ways of lessening the competition. At first, the New River and the West Middlesex tried to merge, but their proposal was rejected by Parliament in 1816 after lobbying from other water companies and parish authorities. A second attempt in 1817 was more successful. Rather than merging, the companies carved up London between them. The New River, the Chelsea, the West Middlesex, and the Grand Junction were all parties to an 1817 agreement governing the west of the metropolis. They agreed to supply only certain zones, and not attempt to venture outside set boundaries, thereby ensuring that only one supplier was available in most of the city. Once the agreement was in place, they abandoned, dug up, or transferred pipes to other companies in streets outside their agreed zones. In the east, the New River had signed another agreement with the East London Waterworks in 1815, allowing it to keep the City, with the East London taking most of the areas farther east. Finally, the Hampstead Waterworks also agreed with the New River on a supply zone demarcation in 1817. Although many of their customers were enraged by this new state of affairs, forming an "Anti-Water Monopoly Association," the agreement largely held for rest of the century. The smaller companies were soon also settled. The York Buildings Company was removed as a competitor when the New River leased its assets in 1819, with its charter revoked not long afterward. The LBWW finally succumbed in 1822. It had not managed to renew its pipes with iron, and the City decided to replace London Bridge. The LBWW was then taken over by the New River.[14]

Legacy of the London Water Network

From around 1800, London's water network—particularly its widespread connections to houses—was an inspiration to urban networks in various ways. This was most evident with other water supply infrastructure. Government authorities when first designing their own integrated systems in other countries either explicitly looked to London as model or relied on English water engineers. As early as 1765, steam engines were attractive to French promoters of water companies.[15] The proposals of the Périer brothers looked to England for their inspiration; their prospectus used London's steam engines as a basis for their own business plan, claiming that some French citizens who had traveled to London became very jealous of the abundant profusion of water supplied in that city.[16] The brothers also traveled to England to study the water supply situation there, the first time in 1777, followed by another in 1778.[17] In 1779 they went again, this time to Birmingham to negotiate a deal with Boulton & Watt for a steam engine of their design.[18]

In 1802 Napoleon approved the construction of an aqueduct from the Ourcq River to Paris, to be distributed mostly from public fountains he hoped would adorn his capital. The project took twenty years to complete. In this case, however, there appeared to be little attempt to emulate the London model, and it was explicitly rejected, including by Louis Bruyère, an engineer originally tasked with preparing a study for the new canal. In his opinion, the New River's open channel made its waters muddy, and its wooden pipes gave it a bad taste.[19] Even after the Ourcq project opened in stages in the 1820s the basic model remained as it had been: public fountains and private wells. Renewed interest in the London model came when the restored monarchy and its successors tried to build up Paris infrastructure further. French engineers made repeated visits to London in the 1820s and 1830s to learn from the city's water system, producing a series of reports about how to replicate it in Paris.[20] Among the first was the engineer and prefect of the Seine département Jacques Chabrol de Volvic in 1823. He became convinced that the best model for Paris was to rely on private companies to supply water because the state would not have the financial resources to complete the project. He was followed in 1824 by the state civil engineer Charles-François Mallet, who was sent to study the English model in greater detail. In the meantime, an English engineer working for the Grand Junction Waterworks named Anderson was brought over to help design the urban distribution system for the newly complete Ourcq project.[21] Mallet returned to Paris and prepared plans for improving the city's water supply that were, in his words, meant to "imitate the one adopted in England, and particularly in London."[22] His system, however, envisaged half going to private supply, with the rest for public fountains. Although his vision for the redesign of the entire system could not be implemented straightaway, he did introduce some techniques he had learned about, such as how to lay pipes more rapidly.[23] Despite their hopes, however, private companies interested in water supply never materialized, in part due to hostility in some quarters to private supply, and the situation in Paris languished until the 1850s. Other engineers visited London at that time to study the situation once again, albeit with more criticisms than Mallet had made, in part due to the problems with filtration that had become painfully evident in the preceding years. They nevertheless recognized that, with 95 percent of houses in the city with water, there was still much to emulate.[24] By the later part of the century, however, Paris had so far surpassed London in quality and volume served that French engineers began to make disdainful comparisons.

The first German city with an integrated water system was Hamburg, and this system too was largely based on the English model. The city had a pumping engine supplying houses from the 1530s. Although the engines were expanded and replaced over the years, the scale of the network had not grown like London's had in the intervening century, with about five hundred taps in the early nineteenth century. Hamburg had not experienced London's population growth and had about 107,000 inhabitants in 1810.[25] The first steps in modernizing the city's system were made in the 1820s when a couple of new water companies were established. A resident of the city, Georg Elert Bieber, created a small company to provide water to parts of the city not served by the old waterworks. His first attempts proved abortive when his buildings were destroyed in 1813 during the Napoleonic Wars, but he set up pumps again in 1822. These were replaced with two Boulton & Watt steam engines in 1832. The numbers supplied, however, remained small; in 1840 the company connected to about 380 taps.[26] A larger company was established in 1831 by an English merchant, Edward James Smith, working with William Chadwell Mylne, the New River's chief engineer. The company's first pumps were consumed by a fire that destroyed much of the city in 1842, but new Cornish steam engines were built on the Elbe, and these were soon supplying about one thousand houses.[27] In the meantime, the three old waterworks merged in 1836 in an attempt to combat declining sales lost to the new companies. The city council had in the meantime been exploring ways to improve supply, but the great fire of 1842 destroyed the old pumps before any action was taken. The city then expropriated the land belonging the old waterworks.[28]

William Lindley (1808–1900) was an English civil engineer who had come to Hamburg in 1834 to build railways with another engineer, Francis Giles. After the fire, he was hired as a consulting engineer for the city's reconstruction, including the city's sewer and water supply systems. He came up with a new city plan and was later involved in the design of gasworks there.[29] He drew a plan for the water supply network with Frank Giles, who worked for the New River Company in London. Frank was Francis's nephew. Lindley also consulted with William Mylne, as Smith had before him, and together they prepared a proposal for the city.[30] Their proposal included two steam engines as well as the dual system of mains and service pipes developed in London the pervious century. Lindley and Mylne also proposed the English system of fireplugs, as well as the use of air plugs to bleed the network when needed.[31] The city council adopted the plan and placed Lindley in charge of executing the project, working with English engineers. The steam engines and many pipes were produced by the Butterley Ironworks in Alfreton, Derbyshire. By the end of 1846, about 750 customers were connected.[32] At this point, the population had reached 137,000.[33]

In the United States, Philadelphia was the first city to build a water distribution network, using technology and expertise imported from England. The city instituted a watering committee to evaluate proposals for the construction of a water supply system and accepted one by the immigrant English engineer and architect Benjamin Latrobe (1764–1820). Latrobe had worked in England as a canal engineer, likely for the engineer William Jessop, as well as an architect and surveyor for the London Police Offices. Latrobe had moved to the United States in 1796 after the death of his wife. He was working as an architect in Virginia when he got the Philadelphia commission in 1799. He recommended using steam engines to pump from the Schuylkill River into a pipe network made from bored logs. The system was functioning by 1801. The engines were made at the Soho plant near New York City, which was the only manufacturer trained in making the Boulton & Watt engine in America. Latrobe's original pumps in Centre Square proved too unreliable and expensive, however, and were replaced by other steam engines on the river by 1815. Philadelphia's experience with waterworks was seminal for similar projects throughout the United States. The city received much publicity and, as a result, many visitors. New York City had tried to use an English engineer to build a waterworks, but the project was aborted by the outbreak of the Revolutionary War. It was only after 1800 that it tried again.[34]

The network model inspired not only other water networks but also other kinds of technological networks, the first of which was gas distribution. The world's first integrated gas distribution network was built in London from 1812 to 1820, and the city water supply was explicitly an inspiration for the new network. When the gas industry was first established in the early nineteenth century, the original model as designed by Boulton & Watt used individual gas plants for each building. These plants would gasify coal, and the gas was distributed to lamps throughout the structure. The model of a citywide centralized distribution network, connecting directly to thousands of houses and streetlights and fed by a few central gas plants, was a later innovation. The man who initially provided this new network vision for the gas industry was Frederick Winsor, a German immigrant and entrepreneur. In his promotional material, he explicitly relied on London's water network, and the New River Company in particular, as a model that the fledgling gas industry could follow. He pointed not only to the integrated system of mains distributing water into houses but also to the business model that water companies had adopted.[35] Gas was to serve in turn to serve as a model for Thomas Edison when he designed his electrical system. Much like Winsor before him, Edison eschewed individual installations for electrical networks and sought to design all the components of a new large-scale electrical network that featured powerful generators in large plants feeding regional distribution networks with thousands of customers.[36]

The construction of integrated urban sewer networks was also linked to water supply, although in a way different from gas and electricity. The purpose of sewers is to collect and remove waste, not to distribute and supply. Furthermore, whereas the for-profit business model was part of what the builders of gas and electrical networks emulated, this was decidedly not the case with sewers, which were built by municipal governments. It was, however, precisely as a complement to water networks that the sanitary reformers of the mid-nineteenth century designed sewer networks. These reformers were primarily concerned with water-borne pollution, and their solution was the continuous circulation of water.[37] Most famous among them was Edwin Chadwick, who in his 1842 report on the conditions of the poor in England, presented a proposal for a hydraulic system that integrated both water supply and sewerage. Lacking any integrated sewer system, most of London's waste accumulated in cesspits, or ran haphazardly into rivers and open sewers, leading to terrible living conditions and frightful epidemics. The solution from the technological perspective at least was proper circulation of water through the city in a hydraulic system. In his report, Chadwick argued that cities where water supply directly connected to houses were healthier: "In Paris and other towns where the middle classes have not the advantage of supplies of water brought into the houses the general habits of household and personal cleanliness are inferior to those of the inhabitants of towns who do enjoy the advantage."[38] He and his engineers, most notably John Roe, elaborated on this integrated system between 1843 and 1845, although, lacking any meaningful legal and administration structure to implement it, they produced little immediate results.[39] It took until the "Great Stink" of 1858 before Parliament approved the construction of a major sewer network to prevent dumping straight into the Thames in London (fig. 8.2). In the meantime, other engineers and sanitary reformers influenced by Chadwick elaborated and implemented integrated sewer networks. These include William Lindley in Hamburg, and F. O. Ward in London, who in 1852 presented a vision for an arterial-venous system of city sanitation in a speech given in Brussels. He argued that sewers were a complement to water supply: "Thus we construct in a town two systems of pipes the one bringing in pure water the other carrying off this water enriched by fertilising matter." The heart of the system was a steam engine that sets the water in motion in the "vast tubular organisation." Just like water networks had done away with tankard bearers, the new system involved the elimination of human labor in cleaning cesspools, and all carriage of waste was to be effected by water flow.[40] In the United States, Chicago and Brooklyn were the first cities to build comprehensive and integrated sewer systems from the 1850s, and they relied on the British model.[41]

Conclusion

The years from 1790 to 1820 were tumultuous for the London water industry as some long-developing trends combined with specific events to create strong pressure for change. The French Wars drove up the price of wood far beyond anything the industry had seen, shattering the wood-pipe model that had dominated since 1580. This combined with the increasing availability of ever-cheaper iron. The existing water companies, firmly ensconced in their dominant positions, were too conservative to adopt iron pipes wholesale. The new entrants, however, were not, enabling them to build large networks rapidly. The ongoing improvement in steam engine efficiency further created opportunities for the new companies to provide water, in some cases at higher pressures and with constant supply. A new joint-stock boom was the final element in the explosive mixture that shook the London water industry, finally ending the New River's rise, which had been running inexorably from 1700, and even as far back as 1650 with the exception of its stumble around 1685. The company lost customers and finally abandoned areas of the city to rivals in districting arrangements. Iron and steam, both parts of the classic

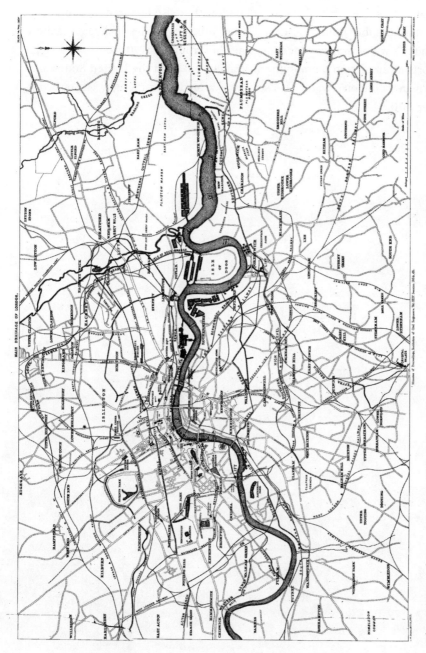

Figure 8.2. London's sewers in 1865. Joseph Bazalgette, "On the Main Drainage of London," *Minutes of Proceedings of the Institution of Civil Engineers* 24 (1865), p. 38

industrial revolution, had had an important impact on the industry. As transformative as these decades were in London, they also featured the industry's exerting notable influence. People from other countries had remarked on London's water industry for some time, but from the 1770s there were clear efforts at emulation, notably from France, and then the United States and Germany. London's water also served as a model for the new gas network after 1810.

Conclusion

The London water supply industry was never a revolutionary phenomenon. At each stage of its history, there was strong and evident continuity with what had preceded it. At its origin around 1580, the model of a for-profit water company providing water to houses incorporated many elements of what came before. The London Bridge Waterworks served a public fountain but also connected to some houses, much as the conduit supply did. Selling water in discrete batches was just what the tankard bearers did, but it was now supplied by pipe rather than person. Likewise, the conduits were not simply displaced. Rather, like the tankard bearers, they faded in importance and disappeared over the decades. Similarly, there never was a clear transition to a different kind of supply network, one that placed London's infrastructure in a distinct category from what was found in German cities. With time, London's system grew, adding many more connections so that by the early eighteenth century, the number of people connected to London's network far surpassed anything found in other cities. The one possible exception to this at this time was Edo, one of the few cities of comparable size to London outside Europe. Its water supply consisted of numerous canals feeding a system of ditches distributing water to communal wells in many parts of the city.[1] Over the course of the eighteenth century, London's system continued to expand, and its engineers maintained its stability in the face of an ever-expanding population, not only keeping up with demand but even increasing the proportion of houses served with direct connections. They adopted more sophisticated systemic solutions, such as a doubled pipe network and centralized control through accurate mapping of the network. These novel techniques meant that the transition in scale to a large network was more than adding to existing infrastructure. It entailed more refined control of assets.

The difference between the New River, on the one hand, and the LBWW, on the other, underscores this. Hearne's report from 1745 indicated that the LBWW in the mid-eighteenth century had relatively few such controls, whereas the New River had been confronted with scale-related problems before 1700 and worked to address them for decades thereafter. The success of these measures became evident. By the mid-eighteenth century, foreigners were remarking on London's supply, and eventually some began to try to import the model. The years around 1800 were perhaps the most revolutionary time for the water industry, when exogenous factors, notably the rapidly rising price of wood, the increasing availability of iron, more efficient steam engines, and a new joint-stock boom all combined to reshape the industry. Even so, many elements proved resilient, including possibly unattractive ones such as intermittent supply.

This long-term growth and stability, which made London's water supply exceptional, did not have a single cause. The question of why some of England's (and Britain's) industries diverged from its continental peers in terms of size and technical development through the classic industrial revolution continues to generate much interest, and the history of the water industry, which has not been evaluated in this context, was one of the industries that clearly displayed a divergence, albeit one dating well before 1760–1830, the approximate traditional dates of the industrial revolution. The exceptional nature of the water industry had various roots and was sustained by different factors over time. Among these was technological innovativeness. Water technologies were not originally English in origin, but as they became indigenized, so too did innovation in the industry. The late seventeenth century marked a notable period of technological creativity, which included the work of George Sorocold with waterwheels and Christopher Wren and George Lowthorp with network issues. The eighteenth century also featured regular innovations, including improved waterwheels with John Smeaton and others; the introduction and improvement of steam engines by Thomas Savery, Thomas Newcomen, Smeaton, and Boulton & Watt; network design; and then large-scale use of iron piping and other components. This technological creativity was quite broad-based both chronologically, in evidence sporadically from 1660, and in different kinds of technologies within and outside the water industry, ranging from materials, to prime movers, to network design. To be sure, the innovative spirit behind these developments was not constantly in evidence within the industry. The water companies' relative conservatism in regard to adopting iron piping was a case in point.

It was only when new entrants came to the market that these were used extensively. Nevertheless, there were regular bouts of technological innovativeness from at least the later seventeenth century allowing the industry to survive and flourish.

What was the root of this ongoing technical inventiveness? It is difficult to point to a single source, if only because not enough details of the origins of the some of these innovations remain. Many of the New River's and the LBWW's design decisions in the eighteenth's century were anonymous. In terms of named leading figures, little is known about George Sorocold's background and education. The same is true of the New River's first chief engineer, Henry Mill. Robert Mylne, who succeeded him, was educated in hydraulics mostly by working in Italy, studying ancient Roman water systems. Samuel Hearne, the LBWW's chief engineer, by his own admission learned on the job. His successor John Foulds was a millwright who also acquired his knowledge of waterworks on the job. Some historians of the industrial revolution, such as Larry Stewart and Margaret Jacob, have claimed that science (or natural philosophy) had an indispensable part to play in Britain's technical creativity as Newtonian culture flourished in the country in the eighteenth century.[2] While it is true Wren and Lowthorp, who did make important contributions, were both members of the Royal Society, it is not evident that their suggestions drew on any knowledge specific to their membership there. Almost all the development apparent throughout the entire history of the water industry was incremental and did not usually involve macro-inventions coming from theoretical research. Even the steam engine, the clearest case of a radically new technology adopted by the industry, came into use slowly over decades and only after gains in efficiency. It was really only in the nineteenth century that it transformed how water was supplied. The water industry was largely sustained by the practical and efficient craftsmen and artisans who learned by being involved in their work in the industry. The best model to describe how the water industry functioned is that of collective invention used by Robert Allen and Alessandro Nuvolari.[3] Using different examples, they argued that technological innovations sometimes occurred as a regular flow of improvements accumulating over time to produce significant results, even if none of the individual steps were notable or patented. The entire history of technological innovation in the water industry was of this sort: slow, individual improvements, most being so small that their sources have long disappeared, but whose accumulation produced a vast, stable, effective technological network that eventually deeply impressed observers.

Business legal models also gave solidity to the London water industry, especially the joint-stock form that, although not entirely new around 1600, enjoyed a burst of popularity. This model was first formally adopted by the New River Company in 1619 when it became a joint-stock corporation, and it was followed many decades later by a number of other companies in the joint-stock boom of the late seventeenth century, including by the LBWW, which transformed into an unincorporated joint-stock company. The primary advantage of the joint-stock form was that it helped to gather the capital required by the larger companies because joint-stock shares were tradable, and its capital was permanent. The New River used the investment of tens of shareholders in its early years to build its infrastructure. The New River in effect carried forward the characteristic of permanent joint-stock capital in the water industry from its founding until the form became popular again and new companies emulated it.[4] The LBWW and the Chelsea Waterworks both did so after 1700 when they expanded their own works. This was true once again with the post-1800 companies that reshaped the industry. There were some partnerships, but these remained small companies. Some business historians have argued that the corporation was a crucial feature of the development of the modern economy.[5] In the case of the water industry, the joint-stock form was indeed central, although being a corporation was less so. The LBWW seemed not to suffer for its unincorporated status.

Governments also had an important role to play. At times, they were seeking unearned income from the water industry, as was the case with the Crown's repeated granting of patents and charters to companies before the Civil Wars without any investment on its part. The City of London did the same at the end of the seventeenth century when, under duress with its bankruptcy, it demanded large fees for granting water rights. In most of these cases, the result was the failure of companies concerned. At other times, governments were willing to invest, such as in the LBWW, the New River, and others among the earliest companies. Indeed, the early companies that became established and survived were precisely those that received some funds, either as loans or through equity from governments. This pattern of direct government intervention diminished with time. The Crown's effective powers withered, especially after 1688, and Parliament largely limited itself to granting acts.

Another important facet of government action in the water industry was support for rights, either for water or for access to land and streets. Governments were willing to grant these rights and uphold them. Entrepreneurs had to negotiate the thickets of conflicting and changing political power bases as the Crown and Parliament sparred before 1688, as well as factions within Parliament. In addition, the City had real power. One of Myddelton's strengths was his astuteness in managing politics. Despite the shifting politics, however, there was only once case of water rights granted—in this case by Oliver Cromwell—and later arbitrarily revoked by an opposing political force. The result was that, although the uncertainties were real enough, the industry managed to hold its acquired rights for hundreds of years. New entrants found, however, that the existing companies and vested interests could use political machinations to stymie their efforts at establishing new companies, as the many failed attempts to bring water from the River Colne showed. The courts, furthermore, upheld these rights. The land access rights granted to the New River Company in its charter and in the parliamentary acts granted to the City were interpreted broadly in a couple of court cases around 1700. These argued that landowners could not arbitrarily remove pipe or raise rents to extortionate levels once leases had expired. The courts did so on the basis of an argument from the great public good the New River provided. In all this, the Glorious Revolution of 1688, which was important for fostering investment in other infrastructure industries, and likely for water supply in other English cities, was not a key turning point for water in London. It did, however, end the uncertainty that had hung over the industry with the political changes, although that was probably not evident until many years had passed. Finally, patents seemed to play little role in the water industry as a means for protecting new inventions and thereby encouraging investment in new technologies. The importance of the English patent system for fostering the industrial revolution remains unclear and subject to debate.[6] Patents were present in the water industry, but before they were specifically associated with new inventions, and were a means for Crown patronage. In the eighteenth century, patents were of no discernible importance for London water, except in their use for steam engines.

Another major support for the London water industry was consumers, who furnished the demand for all those water connections. As the experience of Paris showed, having the water available was not enough. People had to be wealthy enough and sufficiently interested to pay for it. London, as a rapidly expanding, commercial, wealthy city, with rich areas, especially the expanding West End, was fertile ground for water supply. Relatively second rate when the industry was first born, the city's growing population with wealth based in commerce, finance, government, law, and manufacturing, all supported by a large retail and service sector, meant that by 1700 it had reached the top level among European and world cities in wealth and size. Many in the expanding middle and upper classes were willing and eager to be connected to the water supply network. The growth in demand for connections took place especially after 1660 and during the reconstruction of London after the Great Fire, and coincided with the consumer revolution that Jan de Vries has argued was fundamental in establishing the new patterns of consumer behavior that would sustain the long-term economic growth of Britain and eventually its industrialization. Water was not a new consumer commodity, but receiving it via pipes in houses was a new way of consuming it, allowing for greater volumes for domestic uses than what bare biological needs required. These uses included cleaning, fountains, and, later in the eighteenth century, regular bathing and water closets. The deep market penetration of water connections in the eighteenth century, growing from about 45 percent in 1700 to 75 percent in 1800, reveals how much consumer demand sustained the industry's growth. Even before 1700, water connections had become commonplace.

What is the London water industry's relevance for the industrial revolution? On the issue of timing, the traditional dates of the industrial revolution, approximately 1760–1830, were not particularly exceptional. It is true that some of the technologies that have traditionally featured importantly for the industrial revolution, including iron and steam engines, had notable effects on the water industry, and, to this degree at least, it was touched by contemporary exogenous technological change. But within the entire history of the industry to 1820, this period was not very different. Technological creativity was in evidence much earlier. Furthermore, not one single factor seems to have dominated the industry's growth, and many of the possible causes put forward to explain Britain's sustained industrial and technological precocity in the eighteenth century were present: consumption, skilled artisans, strong legal institutions, relatively benign government, and local peace after 1660. Coal had relatively little role to play until iron and steam engines were important in the early nineteenth century, and the London water industry flourished independently of coal to 1770. Finally, the water industry was a large-scale urban network connecting to houses, and it served as precursor to the many that would follow in the nineteenth century.

Number of Houses and Customers Served
by London Water Companies

Year	Company	Customers[a]	Houses	House/customer	Source	Page
1614	NR	360			LR 2/25–43	
1615	NR	384			Tynan 2002	345
1615	NR	384			Rudden 1985	23
1615	NR	765			LR 2/25–43	
1616	NR	1,035			LR 2/25–43	
1617	NR	1,215			LR 2/25–43	
1618	NR	1,521			LR 2/25–43	
1618	NR	1,549			ACC/2558/NR/13/304	36
1619	NR	1,343			LR 2/25–43	
1620	NR	1,499			LR 2/25–43	
1622	NR	1,474			LR 2/25–43	
1623	NR	1,249			LR 2/25–43	
1624	NR	1,419			LR 2/25–43	
1625	NR	1,167			LR 2/25–43	
1626	NR	1,089			LR 2/25–43	
1627	NR	1,514			Jenner 2000	269
1627	NR	1,210			LR 2/25–43	
1628	NR	1,375			LR 2/25–43	
1629	NR	1,523			LR 2/25–43	
1630	NR	1,372			ACC/2558/NR/13/304	36
1630	NR				Ward 2003	58
1638	NR	2,154			Jenner 2000	257
1655	Broken Wharf	600			Jenner 2000	257
1670	NR	8,424	10,951	1.3	BL Harley 3604	
1670	NR	8,424	10,951	1.3	BL Harley 3604	
1671	NR	9,314	12,109	1.3	BL Harley 3604	
1672	NR	9,635	12,526	1.3	BL Harley 3604	
1673	NR	9,887	12,853	1.3	BL Harley 3604	
1674	NR	10,260	13,338	1.3	BL Harley 3604	
1675	NR	10,728	13,946	1.3	BL Harley 3604	
1676	NR	10,946	14,230	1.3	BL Harley 3604	
1677	NR	11,027	14,335	1.3	BL Harley 3604	
1678	Hyde Park		500		van Lieshout 2012	193
1678	NR	11,098	14,428	1.3	BL Harley 3604	
1679	NR	10,726	13,943	1.3	BL Harley 3604	
1680	NR	10,537	13,698	1.3	BL Harley 3604	
1680	NR	10,537	13,698	1.3	BL Harley 3604	

Year	Company	Customers[a]	Houses	House/customer	Source	Page
1681	NR	12,446	16,180	1.3	BL Harley 3604	
1682	NR	13,447	17,481	1.3	BL Harley 3604	
1683	NR	13,966	18,155	1.3	BL Harley 3604	
1683	NR	13,966	18,155	1.3	BL Harley 3604	
1685	All companies		45,849		Hooke and Derham 1726	168
1700	York Buildings		2,700		Tynan 2002	348
1700	York Buildings	2,250	2,700	1.2	Dickinson 1954	49
1708	LBWW	*5,200*	5,564	1.07	NUL Pw2 Hy 362	3
1708	NR	*16,800*	23,520	1.4	NUL Pw2 Hy 362	3
1710	NR	*14,147*	19,806	1.4	ACC/2558/NR/14/001 (06), 1710/1-02-01	
1715	Millbank		1,250		van Lieshout 2012	193
1718	All companies		110,000		G. *Geography epitomiz'd* 1718	41
1720	NR		30,000		Stow and Strype 1720	vol. 1, 26
1720	Shadwell		1,400		ACC/2558/MW/C/15/098	
1724	Shadwell	1,187			van Lieshout 2012	193
1725	Millbank	1,040			ACC/2558/CH/01/002	62
1725	NR	20,000	28,000	1.4	*Observations upon the bill now depending, for supplying the cities of London and Westminster, and places adjacent, with water* 1725	
1725	Shadwell	1,120			ACC/2558/MW/C/15/098	
1727	Chelsea	727	800	1.1	ACC/2558/CH/01/002	205
1729	Chelsea	2,560	2,816	1.1	ACC/2558/CH/01/003	66
1730	Chelsea	2,622	2,885	1.1	ACC/2558/CH/01/004	95
1732	Chelsea	*3,400*	3,740	1.1	ACC/2558/CH/01/005	137
1733	NR	21,429	30,000	1.4	Seymour and Stow 1733	26
1737	All companies		85,805		Price 1780	1–2
1737	All companies		85,805		Price 1773	182
1737	All companies		85,805		Maitland 1739	519
1737	All companies		95,958		Maitland 1739	532
1738	Chelsea	*3,750*	4,125	1.1	ACC/2558/CH/01/008	122
1739	NR	21,429	30,000	1.4	Maitland 1739	630
1745	LBWW		8,000	1.093	ACC/2558/MW/C/15/102	10, 43
1745	LBWW	7,320			ACC/2558/MW/C/15/102	10, 43
1751	NR		40,000	1.333	"Sir Hugh Middleton's Scheme for supplying the City of London with good and wholesome Water" 1751	312

Year	Company	Customers[a]	Houses	House/customer	Source	Page
1752	NR	30,000			ACC/2558/NR/05/066, #5	5
1757	All companies		87,614		Price 1780	1–2
1759	NR		40,000	1.527	B. Martin 1759	239
1769	NR	26,197			ACC/2558/NR/12/001	
1774	NR		40,000		*A description of the county of Middlesex 1775*	147
1775	NR	28,571	40,000	1.4	*A description of the county of Middlesex 1775*	147
1775	Shadwell		8,000		Matthews 1835	112
1775	York Buildings				Wicksteed 1835	11
1776	NR		30,000+			
1777	All companies		90,570		Price 1780	1–2
1779	LBWW	8,805	9,510	1.08	ACC/2558/LB/01/002	15
1782	NR	32,143	45,000	1.4	Volkmann 1782	424
1790	NR	34,151	47,811	1.4	ACC/2558/NR/14/002, 1814-03	
1794	LBWW	9,715	10,492	1.08	ACC/2558/LB/01/004	
1795	Shadwell		8,000		Lysons 1795	vol. 3, 383–90
1797	Lambeth		3,000		ACC/2558/LB/01/002, 1797-10-07	107
1800	Chelsea	9,500			Ward 2003	161
1800	LBWW	10,000			Ward 2003	161
1800	NR	59,000			Ward 2003	161
1800	NR	37,472	52,460	1.4	ACC/2558/NR/14/002, 1814-03	
1800	NR		55,000		Rudden 1985	98
1800	York Buildings	2,250			Ward 2003	161
1801	All companies		121,189		Marshall 1832	3
1801	All companies		142,042		*Acts of the Privy Council* 1896	
1804	Chelsea	6,480	8,424	1.3	Minutes of evidence (706) 1821	245
1804	NR		54,681	1.424	Minutes of evidence (706) 1821	207
1804	NR	38,403			Minutes of evidence (706) 1821	207
1804	Shadwell		61,257		Minutes of evidence (706) 1821	245
1804	York Buildings		2,000		Sisley 1899	9
1804	York Buildings		2,089		Minutes of evidence (706) 1821	245
1808	NR	42,960	59,098	7.483	"London" 1829	181
1809	Chelsea	7,898	9,477	1.2	Minutes of evidence (706) 1821	245
1809	East London		10,739		Minutes of evidence (706) 1821	228

Year	Company	Customers[a]	Houses	House/customer	Source	Page
1809	LBWW	9,733	10,317	1.06	Minutes of evidence (706) 1821	245
1809	NR		59,058	1.375	Minutes of evidence (706) 1821	207
1809	NR	42,960	59,098	2.00	Minutes of evidence (706) 1821	207
1809	York Buildings	2,217	2,250		Minutes of evidence (706) 1821	24, 247
1810	NR	45,762	61,779	1.35	ACC/2558/ NR/14/002, 1814-03	3
1810	NR	44,264	59,757	1.35	ACC/2558/NR/14/002	10
1810	Shadwell		10,739		Minutes of evidence (706) 1821	245
1810	York Buildings	2,636			Wicksteed 1835	11
1811	All companies		142,320		Marshall 1832	3
1811	All companies		166,942		*Acts of the Privy Council* 1896	
1811	NR	11,024	14,882	1.35	ACC/2558/NR/14/002	7
1812	East London		18,975		Minutes of evidence (706) 1821	228
1812	NR	42,070	56,794	1.35	ACC/2558/NR/14/002	7
1813	NR	40,847	55,143	1.35	ACC/2558/NR/14/002	10
1814	Chelsea		9,862		Minutes of evidence (706) 1821	217
1814	East London		23,250		Minutes of evidence (706) 1821	228
1814	Grand Junction		1,558		Minutes of evidence (706) 1821	240
1814	NR	40,391	54,247	1.34	Minutes of evidence (706) 1821	207
1814	York Buildings		2,740		Minutes of evidence (706) 1821	225
1816	East London		27,731		Minutes of evidence (706) 1821	228
1816	Grand Junction		2,784		Minutes of evidence (706) 1821	240
1817	LBWW		10,417	1.042	Dickinson 1954.	58
1818	York Buildings		2,636		Minutes of evidence (706) 1821	225
1819	Chelsea		8,632		Minutes of evidence (706) 1821	217
1819	East London		29,926		Minutes of evidence (706) 1821	228
1819	Grand Junction		7,180		Minutes of evidence (706) 1821	240
1819	NR	38,406	51,233	1.33	Minutes of evidence (706) 1821	207
1820	Chelsea	7,193	8,631	1.2	Peppercorne 1840	65
1820	East London		32,071		Peppercorne 1840	65
1820	Grand Junction		8,000		Dickinson 1954	100
1820	Grand Junction		7,180		Peppercorne 1840	65

Year	Company	Customers[a]	Houses	House/customer	Source	Page
1820	Lambeth		11,487		Minutes of evidence (706) 1821	68
1820	Lambeth		11,487		Peppercorne 1840	65
1820	NR	40,063	52,082	1.3	Peppercorne 1840	65
1820	South London		5,200		Minutes of evidence (706) 1821	64
1820	Vauxhall		5,200		Peppercorne 1840	65
1820	West Middlesex		10,350		Peppercorne 1840	65
1820	West Middlesex		11,155		Minutes of evidence (706) 1821	234
1821	All companies		164,948		Marshall 1832	3
1821	All companies		199,153		*Acts of the Privy Council 1896*	
1822	LBWW	10,417			Wicksteed 1835	7
1827	NR		70,000	5.645	Dickinson 1954	58
1828	Chelsea	12,400			*Report of the Commissioners Appointed by His Majesty to inquire into the State of the Supply of Vater in The Metropolis, Dated April 21st 1828 1828*	5
1828	NR	66,500			*Report of the Commissioners Appointed by His Majesty to inquire into the State of the Supply of Vater in The Metropolis, Dated April 21st 1828 1828.*	4
1834	NR		73,212		Matthews 1835	449
1835	Chelsea		13,000		Dickinson 1954	58
1835	East London	45,000			Dickinson 1954	58

[a]Numbers are given in italics when they are estimates based on revenue figures rather than numbers quoted directly by the source.

New River Company Share Sales

	Share price (£)	Source	Page
1622	28	Estimate from William Pitt, Gough 1964	77
1626	76.6	Estimate from NA LR 2/41, in Rudden 1985	19
1650?	500	NA C10/188/36 [*Roger Lukyn v. Governor and others of New River Company*] Sold by Hugh Myddelton I to Roger Lukyn	
1654	500	NA C10/188/36 [*Roger Lukyn v. Governor and others of New River Company*] Second sold by Hugh Myddelton II to Roger Lukyn	
1657	500	Rudden 1985	306–7
1698	4,000	*A true copy of several affidavits and other proofs of the largeness and richness of the mines* 1698	
1698	4,194	Waller 1698	22
1700	5,725	BL Add MS 70341	4
1707	3,900	Rudden 1985.	306–7
1708	4,725	Hatton 1708	vol. 2, 792
1713	7,000	Rudden 1985	306–7
1732	5,000	Rudden 1985	306–7
1740	5,750	NUL Pl F3/4/2 (date range 1737–42)	
1743	4,500	Salmon 1743	297
1744	5,000	Horseman 1744	vol. 1, 464
1745	5,000	*The curious modern Traveller* 1745	
1751	5,250	"Sir Hugh Middleton's Scheme for supplying the City of London with good and wholesome Water" 1751	311
1765	8,000	*London Evening Post*, July 30, 1765	
1766	8,800	*London Evening Post*, February 13, 1766	
1767	9,000	*Gazetteer and New Daily Advertiser*, February 27, 1767	
1771	7,050	*Derby Mercury*, July 19, 1771	
1775	6,923	Rudden 1985	306–7
1782	8,000	Volkmann 1782	vol. 2, 425
1784	7,000	*St. James's Chronicle or the British Evening Post*, October 5, 1784	
1792	15,060	*Diary or Woodfall's Register*, February 29, 1792	
1795	14,000	*True Briton*, December 31, 1795	
1805	15,840	Hughson 1809	22
1808	18,000	*Manchester Mercury*, September 13, 1808	
1821	8,740	Rudden 1985	306–7

Abbreviations

BL	British Library
C	Chancery (NA)
CSPD	Calendar of State Papers Domestic
JHC	*Journal of the House of Commons*
JHL	*Journal of the House of Lords*
Jour.	Corporation of London, Journal of the Common Council (LMA)
LMA	London Metropolitan Archives
MB	New River Company, Minutes of the court of directors (LMA)
NA	National Archives (UK)
NR	New River Company
NUL	Nottingham University Archives
PC	Privy Council (NA)
PP	Parliamentary Papers
RCHM	Royal Commission on Historical Manuscripts
Remem.	City of London correspondence (Remembrancia) as found in Overall 1878
Rep.	Corporation of London, Repertory of the Court Aldermen (LMA)
RLW	Report on the State of the London Bridge Waterworks, 1745. Written by Samuel Hearne. LMA ACC/2558/MW/C/15/102
SP	State Papers (NA)
T	Treasury Board (NA)

Introduction

1. J. Ward 1974. Szostak 1991. Albert, Freeman, and Aldcroft 1983.
2. Tomory 2012. Chandler 1977. Millward 2005.
3. Chandler 1977.
4. De Vries 1984. Malanima 2010.
5. Earle 2001.
6. Slack 2014, 12.
7. North and Weingast 1989.
8. Bogart 2011.

9. Slack 2014.

10. Mokyr 2009. Mokyr 2002.

Chapter One · The Roots of a New Water Industry

1. Knissel and Fleisch 2004, 4–5.

2. Moeck-Schlömer 1998, 165–253.

3. H. Turner 2013, 161.

4. Thirsk 1978, 60.

5. Sheppard 1998, 132.

6. Boulton 2000, 332–34.

7. *London lives, 1690 to 1800* 2015.

8. Innes 2001, 65–67.

9. Sheppard 1998, 142–47.

10. Sheppard 1998, 135–38.

11. Beier 1986, 147–48.

12. Earle 2001, 92.

13. Earle 1989.

14. Clifford 1885, vol. 2, 44, 61.

15. Magnusson 2001. Bond 1993. Jenner 2000. Clifford 1885, vol. 2, 35–198. For other European cities, see Squatriti 2000. Kucher 2004. Grewe 1991, 53–70.

16. Clifford 1885, vol. 2, 39. Oath made by the Keeper of the Conduit in Chepe (1310-10-31), Riley 1868, 77–78.

17. Riley 1868, 225. Alsford 2011. Keene 2001b.

18. Clifford 1885, vol. 2, 39–40. Ordinance that Brewers shall not waste the water of the Conduit in Chepe (1345-07-18), Riley 1868, 225. Alsford 2011. Such direct connections existed in other medieval and early modern cities: Grewe 1991, 58.

19. Lord Cobham requested a connection from Ludgate Conduit. See Remem. I. 656, p. 554 (1592-04-23). Another request for connection from Lord Cobham. See Remem. II. 32, p. 554 and Remem. II. 321, p. 554 (1594-09-30). All Remem. documents used here are printed in Overall 1878.

20. Clifford 1885, vol. 2, 48.

21. Clifford 1885, vol. 2, 60–64. See also Jenner 2000.

22. Grewe 1991.

23. Keene 2001b.

24. Clifford 1885, vol. 2, 48, 52. "Memoirs of Mr. William Lamb, Founder of Lamb's Chapel, etc" 1783. Stow and Strype 1720, vol. 2, 24–25. Rep. vol. 19, p. 178 (1575/6). In the notes, references to the Rep. will be given as Rep. volume, page (<yyyy-mm-dd> if available). In England, New Year's Day was March 25 until 1752. For this reason, two years are sometimes given for dates between January 1 and March 25 before 1752, one corresponding to the year then in use, and the second for the modern Gregorian convention, although even in contemporary documents, both years are given. Furthermore, before 1752 England used the Julian calendar, meaning that dates were behind Gregorian ones by eleven days for dates after 1700 and by ten days before. The dates cited here are the contemporary Julian dates found in the archives and have not been corrected for modern usage.

25. Robins 1946, 132. See also Stow and Munday 1633, 8–12. On Stow, see Merritt 2001.

26. Robins 1946, 130-32, 134. Alsford 2011.

27. Long 2008.

28. T. Reynolds 1983, 136.

29. Grewe 2000, 151–59. Grewe 1991, 61–70. See also T. Reynolds 1983, 64–70.

30. Hill 1984, 143. N. Smith 1976, 97.

31. Magnusson 2001, 169. T. Reynolds 1983, 77–79; on uses around 1500, see p. 136; on German mining, see p. 140. See also Liessmann 1992, 82–85.

32. Magnusson 2001, 76. Squatriti 2000, 202. See also Knissel and Fleisch 2004, 4–5.

33. Grewe 1991; on cities, see pp. 53–70; on wood pipes, see pp. 32–40. Grewe 2000, 146, 147, 149. Schnitter 1992, 56, on wood pipes in Switzerland from 1350.

34. Grewe 1991, 65–67. Grewe 2000, 156–57.

35. Grewe 1991, 70.

36. Moeck-Schlömer 1998, 271ff.

37. Meng 1993, 8–35. Moeck-Schlömer 1998, 271, for the length around fountains.

38. Bernardoni 2009, 300–302. The first artillery boring machine to be depicted in an engraving was by Biringuccio in 1530. Late medieval machines were depicted before 1500, such as by Taccola around 1470. Bernardoni 2008, 208.

39. Stow and Munday 1633, 9–10.

40. Richards, Payne, and Soper 1899, 3–4. Stow and Munday 1633, 11. Clifford 1885, vol. 2, 45–46.

41. An Act concerning the repairing, making, and amending of the conduits in London 1543. Clifford 1885, vol. 2, 48–51.

42. Clifford 1885, vol. 2, 51. Stow and Munday 1633, 9–10. Sharpe 1894, 18–19. 35 Henry VIII, c. 10. Another act was passed for the River Lea in 1570: 13 Elizabeth c. 18.

43. *Tijdschrift voor geschiedenis, oudheden en statistiek van Utrecht* 1841, 387. *Register van Holland en Westvriesland* 1587, 660.

44. Duffy 1979, 78–79. Colvin 1982, 410.

45. Barber 1992, 59. Colvin 1982, 410.

46. Rep. 22, p. 196 (1590-07-16), p. 199, p. 232 (1590-12-01), p. 257 (1590/1-03-04), p. 276 (1591-05-06), p. 320 (1591-10-15), p. 376 (1591-05-21). Rep. 23, p. 440 (1595-09-21).

47. Alsford 2011.

48. Mossoff 2000. Bottomley 2014.

49. Bottomley 2014, 30–39. MacLeod 2009, 41–43.

50. H. Turner 2013, 155.

51. Cramsie 2000.

52. Mossoff 2000. Macleod 1986. MacLeod 2009, 43.

53. On Morris, see Jenner 2000, 256. Dickinson 1954, 21–22. Overall 1878, 550–51, 553. Sharpe 1894, 19. Ward 2003, 17. Robins 1946, 155.

54. *A catalogue of the Harleian collection of manuscripts* 1759, No. 2093, 123.

55. Barber 1992, 59. Colvin 1982, 410.

56. Remem. I. 28, pp. 550–51 (1580-05-26). Petition of Peter Morrice to Mr. Walsyngham, NA SP 12/106/1 f.131, 1575. Jenkins 1895, citing Patent Roll, 20 Eliz., p. 10, memb. 34.

57. Petition of Peter Morrice to Mr. Walsyngham, NA SP 12/106/1 f.131, 1575. Jenkins 1895.

58. *Acts of the Privy Council of England. New Series* 1896, vol. 12, 69, 1580.

59. Rep. 18, pp. 307, 375 (1574-04-20), p. 377 (1574-04-25), p. 382.

60. Rep. 19, p. 99 (1575-07-15), p. 116 (1575-09-10), p. 188 (1575/6-03-25), p. 210 (1576-06-15).

61. Remem. I. 102, p. 551 (1580-07-15). *Acts of the Privy Council of England. New Series* 1896, vol. 12, 85, 1580.

62. For the negotiations in 1581, see Rep. 20, pp. 75, 83, 95, 97, 102, 120, 144, 167, 168, 170. The indenture approved by the City, Jour. 21, p. 251 (1581-11-24). Sharpe 1894, 19. Remem. I, 102, p. 551, 1580-07-15.

63. Masters 1984, vol. 20, 30–62, #137c. Holinshed 1587, 1348.

64. Stow and Strype 1720, vol. 1, 27. See also Holinshed 1587, 1348.

65. Rep. 20, p. 345 (1582-07-10), p. 346* (1582-07-12), p. 368 (1582-10-15), p. 375 (1582-12-13). Details of lease p. 382 (1582-12-06). Copy of lease agreements in LMA ACC/2558/MW/C/15/222/2.

66. Holinshed 1587, 1348. Stow and Kingsford 1603 [1908], 187–200. Thornbury and Walford 1878, vol. 2, 170–83.

67. Taylor and Trentmann 2011, 199–200. Tynan 2002, 356.

68. Rep. 21, p. 45 (1584-04-02).

69. Rep. 21, p. 72 (1584-07-15), p. 75 (1584-07-21).

70. Holinshed 1587, 1348. Stow and Kingsford 1603 [1908], 187–200.

71. Stow and Munday 1633, 8.

72. Jour. 21, p. 252 (1581-11-24).

73. Rep. 21, p. 360 (1586-11).

74. Rep. 20, p. 170 (1580/1-02-16), p. 303 (1581/2-03-20), p. 409 (1582/3-03-07). Masters 1984, vol. 20, 30–62, #138f.

75. Rep. 20, p. 406 (1582/3-03-05).

76. Remem. I. 449, p. 553 (1582-12). Holinshed 1587, 1348.

77. Clifford 1885, vol. 2, 53. Remem. I. 449, p. 553 (1582-12).

78. Rep. 21, p. 45 (1584-04-02), p. 151 (1585).

79. Rep. 21, p. 353 (1586-10-25).

80. Rep. 21, p. 456 (1587-07-15), p. 467 (1587-09-20?).

81. Rep. 21, p. 576 (1588-07-25).

82. NA SP 29/229 f.205, 1667.

83. Rep. 22, p. 22 (1588/9-01-30).

84. Rep. 21, p. 576 (1588-07-25).

85. Rep. 26, part I, p. 40 (1602-10-21), p. 97 (1602/3-02-03). Science Museum Archives, Berry 1-1, p. 34. PC 2/28, p. 285 (1616-06-02).

86. Jenkins 1911. Robertson 1934. Baldwin 2002.

87. Rep. 22, p. 435 (1592-09-29).

88. Rep. 23, p. 56 (1593-04-25), p. 58 (1593-05-?). Jour. 23, p. 189 (1593-04-24).

89. Rep. 23, p. 62 (1593-05-21), p. 66 (1593-05-31). Jour. 23, p. 196 (1593-05-22).

90. Rep. 23, p. 218 (1594-05-06), p. 237 (1594-05-?), p. 245 (1594-06-20), p. 310 (1594-11-05), p. 313 (1594-11-11).

91. Lease agreement, LMA ACC/2558/MW/C/15/173, 1594.

92. Stow and Munday 1633, 12. Maitland 1739, 167.

93. Jour. 23, p. 270 (1594-05-17), p. 275 (1594-05-29), p. 286 (1594-07-22), p. 297 (1594-07-27).

94. Rep. 25, p. 381 (1602-04-26), p. 403 (1602-05-??). Rep. 26, p. 56 (1602-12-29), p. 259 (1602/3-01-11), p. 295 (1602/3-?), p. 357.

95. Maitland 1739, 167.

96. Rep. 26, p. 259 (1602/3-01-11), p. 295 (1604-?), p. 357 (1604-05-??). Jour. 26, p. 242 (1604-08-22). Lease with Thomas Parradine 1604, LMA ACC/2558/MW/SU/01/040/ 2050.

97. Rep. 28, p. 120 (1607-11-14), p. 249 (1608-07-14).

98. Hatton 1708, vol. 2, 791.

99. Sharpe 1894, 20. Jour. 23, pp. 209–10 (1593-07-19).

100. From John Darge. Berry 1956, 18. NA LR 2/28, 2nd book.

101. Wallis 1976. Skempton 2002.

102. Rep. 23, p. 473 (1595-11-12), p. 536 (1596-05-21).

103. Jour. 25, p. 180 (1600-07-02). Rep. 25, p. 123 (1600-07-21), p. 130 (1600-08-02).

104. Rep. 25, p. 340 (1601-02), p. 379 (1602-04-21).

105. Rep. 26, p. 40 (1602-10-21). The only further mention of Wright is Rep. 29, p. 49 (1609-07-17).

106. NA PC 2/47, p. 321/156 (1637-04-23). Faulkner 1820, 419–20.

Chapter Two · The Birth of the New River Company

1. Rudden 1985, 17. The legal name of the New River as specified in its charter was "The Governor and Company of the New River brought from Chadwell and Amwell to London." Throughout this book, the company is referred to as the "New River Company" or the "New River," as it was usually called throughout its history.

2. History of the New River 1844, LMA ACC/2558/NR/13/304, pp. 26–28.

3. Winfield 2009, 5.

4. Bogart 2011, 1075. I have used £7,500, corrected for inflation from 1750.

5. Its charter lists 29. Rudden 1985, 282.

6. Rosenberg and Birdzell 1986. Micklethwait and Wooldridge 2003. Landes 1969.

7. 13 Geo. I (1727) c. 206. 6 Geo. II (1732) c. 9.

8. Harris 2000, 94–98.

9. Harris 2000, 16–19.

10. See Harris 2000. Donaldson 1982, ch. 1.

11. Stern 2013. Harris 2000, 52.

12. Harris 2000, 50–52.

13. For a detailed description of the unique legal characteristics of the New River, see Rudden 1985.

14. Scott 1910, vol. 3, 462–70. Harris 2000, 40–45.

15. Freeman, Pearson, and Taylor 2012, 53–57.

16. Freeman, Pearson, and Taylor 2012, ch. 2.

17. Freeman, Pearson, and Taylor 2012, 22. For water in particular, see Hassan 1985.

18. Bogart 2011, 1081.

19. Edmund Colthurst, Licence Under Great Seal Of James I to Cut The New River 1604. LMA ACC/2558/MW/C/15/017. RCHM 1910, 242. RCHM 1933b, 55. CSPD 1603–10, p. 93 (1604-04-11). CSPD volumes will be referred to by year range. Rep. 25, p. 325 (1601-01-19). Jour. 26, p. 322 (1604/5-03-18). See also Rudden 1985, 9.

20. Letters patent dated 1604-04-18, cited in Rudden 1985, 9.

21. LMA ACC/2558/MW/C/15/017, cited in R. Ward 2003, 20.

22. Edmund Colthurst to Lord Cecil, RCHM 1933a, 55, 93 (1604-04-11[?]). Gough 1964, 29. RCHM 1938, 18 (1605-05-04). See also Jour. 26, p. 322 (1604-03-18). Jour. 26, p. 392 (1605-10-22). JHC vol. 1, p. 262 (1605/6-01-31).

23. CSPD 1603–10, p. 93 (1604-04-11). RCHM 1933a, 417. Gough 1964, 29.

24. 3 James I c. 1. 4 James I c. 12. Reprinted in Rudden 1985, 251–55.

25. Sharpe 1894, 20. NR Letters ca. 1745, LMA ACC/2558/NR/13/047/002, p. 2.

26. Rep. 28, p. 288 (1608-10-28), p. 354 (1608/9-02-15). Rep. 29, p. 3 (1609-03-14). JHC vol. 1, p. 311 (1606-05-22). JHL vol. 2, p. 442 (1606-05-26).

27. Rep. 27, p. 235 (1606-07-08), p. 257 (1606-08-25), p. 263 (1606-09), p. 265 (1606-09-11), p. 312 (1606-12-04), p. 325 (1606/7-01-11). Rep. 28, p. 288 (1608-10-28), p. 354 (1608/9-02-15). Rep. 29, p. 3 (1609-03-14).

28. Jour. 27, pp. 377–78 (1609-03-28). Remem. II. 347, pp. 554–55 (1609-07-10). Remem. IV. 46, pp. 556–57 (1616-12-23). NR Letter ca. 1745, LMA ACC/2558/NR/13/047/002, p. 2.

29. Berry 1956. Gough 1964.

30. JHC vol. 1, p. 262 (1605/6-01-31).

31. Rudden 1985.

32. S. Reynolds 2010, ch. 3.

33. Edmund Colthurst to Viscount Cranborne, RCHM 1938, 181 (1605-05-04).

34. Rep. 31, p. 396 (1614-09-06). LMA ACC/2558/NR/13/007, folder 3, p. 164 (1656).

35. RCHM 1872, 58.

36. Rudden 1985, 281. NA LR2/27, p. 1 (1608/9-02-20). Gough 1964, 36.

37. Share indenture, Rudden 1985, appendix E (1612-05-08).

38. Agreement with Sir Henry Neville to sell shares. (1612-05-08), cited in Rudden 1985, 274–78. See also charter Rudden 1985, 279–91.

39. See the entries in Thrush and Ferris 2010. www.historyofparliamentonline.org /volume/1604-1629/.

40. Thrush and Ferris 2010.

41. CSPD 1611–18, p. 106 (1611-??). Scott 1910, vol. 3, 19. Rep. 29, p. 206 (1610-04-17), p. 231 (1610-05-24). Gough 1964, 40–43.

42. Jour. 28, p. 176 (1611-02-27). Rep. 30, p. 100 (1611-03-28). Remem. II. 347, pp. 554–55 (1609-07-10). LMA ACC/2558/NR/13/007, folder 3, p. 158a. The text of the indenture is printed in Rudden 1985, 256–64. Original agreement, Jour. 27, pp. 377–78 (1609-03-28).

43. CSPD 1611–18, p. 128 (1612-02-02). NA, SP 14/141, f.51 (1612-05-02). Text of agreement reprinted in Rudden 1985, 268–73.

44. Remem. IV. 46, pp. 556–57 (1616-12-23). Rudden 1985, 268–73.

45. NR miscellaneous correspondence and papers, LMA ACC/2558/NR/13/13/1. R. Ward 2003, 49.

46. Gough 1964, 74. R. Ward 2003, 46. Approval (1669-11-09), RCHM 1894, 573.

47. Robins 1946, 159. R. Ward 2003, 54.

48. Rep. 32, p. 72 (1614-03-02).

49. The original tenant lists and other documents from the first decades of the New River's existence are preserved in the National Archives, LR 2/27–41. Tenant list from LR 2/35 lists about 270.

50. Overall 1878, 557. Remem. IV. 46, p 556 (1616-12-23). The king gave a second sim-

ilar order in 1622. Gough 1964, 76. RCHM 1858, 409. The company petitioned for further support in 1623: Transcripts of minutes, LMA ACC/2558/NR/13/007, folder 2, pp. 64, 65.

51. NA PC 2/29, p. 404 (1618-06-04).

52. For 1618 to 1626, see collectors' books, NA LR 2/35-42. History of the New River (1844), LMA ACC/2558/NR/13/304, pp. 34–36. For 1627, 1217 tenants, see NA LR 2/43. For 1630 (1372 tenants), LMA ACC/2558/NR/13/304, pp. 34–36. NA LR2/43. For 1638, see Jenner 2000, 257. Records of the Exchequer, NA E 178/6032 m. 10.

53. Slack 1986, 62.

54. NR Letters patent of 21 June 1619, LMA ACC/2558/NR/13/035/3 (1619-06-21). Text at Rudden 1985, 279–91.

55. Scott 1910, vol. 3, 22. William III did invest in the new East India Company in the 1690s. Stern 2013, 180.

56. Rudden 1985, 27.

57. Rudden 1985, 289. Sisley 1899, 161–71, citing Patent Roll (NA), 17 James I, part xvi, no. 67 (1619-06-21).

58. Mossoff 2000, 1269–70. NA SP 29/188, f.217 (1666-01-18).

59. Remem. IV. 46, pp. 556–57 (1616-12-23). Rep. 33, p. 32 (1616-12-25), p. 110 (1617-05-22). Acts of the Privy Council 1927, vol. 35, pp. 99–100 (1616-12-22).

60. Remem. IV. 96 and 97, pp. 557–58 (1617-01). Remem. IV. 101, p. 558 (1617-02-27). NA PC 2/29, p. 286 (1617-02-23).

61. Remem. IV. 101, p. 558 (1617-02-27).

62. RCHM 1874, 283.

63. CSPD 1619–23, p. 409 (1622-06-22).

64. Scott 1910, vol. 3, 23. JHC vol. 1, 611 (1621-05-07). JHC vol. 1, 727 (1623-03-04). JHC vol. 2, 554 (1642-05-03).

65. Harris 2000, 46–52.

66. Gough 1964, 76, citing BL Add MS 29974A, f. 63.

67. Scott 1910, vol. 3, 23–24.

68. NR Transcript minutes Nov. 1619–Sept. 1692, LMA ACC/2558/NR/13/007, folder 1 (1631-11-12). LMA ACC/2558/NR/13/007, folder 2, p. 32a. LMA ACC/2558/NR/13/047/002, p. 3. CSPD, 1631-33, p. 182 (1631-11-15).

69. Sanderson 1732, vol. 19, 244–50. CSPD 1629–31, pp. 553, 555 (1631/2-02-11).

70. CSPD 1637–38, p. 369 (1638-04-17), p. 610–11 (1638-08). CSPD 1639, p. 481 (1639-09-06), p. 123 (1639-11-26). CSPD 1640, p. 438 (1640-02-09).

71. NR Transcript minutes Nov. 1619–Sept. 1692, LMA ACC/2558/NR/13/007, folder 2, pp. 100, 152.

72. Ford and Roberts 1641. JHL vol. 4, pp. 148, 153-54 (1641-02-06). RCHM 1874, 45 (1640/1-01-30). Jenkins 1928. Gough 1964, 78–80.

73. Sharpe 1894, 26, citing Rep. 47, pp. 45, 58, 89, 105, 300.

74. Gough 1964, 64, derived from New River records in the National Archives, LR 2/27–44.

75. Stern 2013, 178.

Chapter Three · Water in the Age of Revolutions, 1625–1730

1. Pettigrew 2011, 25–27.

2. Harris 1994.

3. Rep. 38, p. 12 (1623-11-03).

4. Gough 1964, 77, citing BL Add MS 29974A, f. 64.

5. NR Transcript minutes Nov. 1619–Sept. 1692, LMA ACC/2558/NR/13/007, folder 1, pp. 18, 65 (1624-02-04).

6. NR Transcript minutes Nov. 1619–Sept. 1692, LMA ACC/2558/NR/13/007, folder 1 pp. 23, 71 (1628).

7. NR Transcript minutes Nov. 1619–Sept. 1692, LMA ACC/2558/NR/13/007, folder 1, pp. 36a, 74 (1632-02-08).

8. Remem. VII. 87, p. 559 (1633-03-02).

9. NR Transcript minutes Nov. 1619–Sept. 1692, LMA ACC/2558/NR/13/007, folder 2, p. 115 (1634-11-04).

10. NR Transcript minutes Nov. 1619–Sept. 1692, LMA ACC/2558/NR/13/007, folder 2, p. 115 (1634-11-04), p. 133a (1635), pp. 92, 116 (1635-06-13).

11. R. Ward 2003, 54.

12. NR Transcript minutes Nov. 1619–Sept. 1692, LMA ACC/2558/NR/13/007, folder 2, p. 98 (1637-02), p. 136 (1638-04).

13. Summerson and Colvin 2003, 12–13. Sheppard 1998, 176.

14. NR Transcript minutes Nov. 1619–Sept. 1692, LMA ACC/2558/NR/13/007, folder 2, pp. 87, 102, 145 (1639). Sheppard 1998, 190.

15. Sanderson and Rymer 1732, 245.

16. NA PC 2/47, p. 321/156 (1637-04-23).

17. CSPD 1627–28, pp. 114–15 (1627-03-31).

18. Sanderson and Rymer 1732, 242–450.

19. JHL vol. 4, p. 153 (1640-02-06).

20. Roberts 1641a, 2.

21. RCHM 1874, 94 (1641-08-05). JHL vol. 4, p. 342 (1641-08-05). RCHM 1877, ix (1644-07-19). JHL vol. 6, p. 640 (1644-07-19).

22. Aubrey and Clark 1898, vol. 1, p. 255. JHL vol. 4, p. 148 (1641-01-30), pp. 153–54 (1641-02-06). JHC vol. 2, p. 161 (1641-05-29). CSPD 1637–38, p. 369 (1638-04-17). CSPD 1639, p. 481 (1639-09-06). CSPD 1639–40, p. 123 (1639-11-26), p. 438 (1639/40-02-09). Ford and Roberts 1641. NR Transcript minutes Nov. 1619–Sept. 1692, LMA ACC/2558/NR/13/007, folder 2, pp. 100, 152 (1640). Jenkins 1928. Dickinson 1954, 43–45. Gough 1964, 79–81.

23. Sheppard 1998, 162.

24. NA SP 29/188 f.217 (1666/7-01-18).

25. Dodd 2014.

26. R. Ward 2003, ch. 9.

27. NA SP 29/188 f.217 (1667-01-18). NR Transcript minutes Nov. 1619–Sept. 1692, LMA ACC/2558/NR/13/007, folder 3, p. 170 (1647).

28. Jour. 41, p. 101 (1654-08-02). See also pp. 107, 117, 118, 127, 129, 135. Rep. 63, p. 140 (1654-07). See also p. 437.

29. Jenkins 1928.

30. Ford and Roberts 1641.

31. D'Acres 1660, 17. Jenkins 1928. Rep. 64, pp. 12, 14, 41 (1655-11). Monconys 1665, vol. 2, 29–30.

32. CSPD 1663–64, p. 655 (1664-07-31). CSPD 1664–65, p. 72 (1664-11-16). Aubrey and Clark 1898, vol. 1, 255.

33. CSPD 1664–65, p. 72 (1664-11-16), p. 230 (1664/5-02).

34. Howell 1657, 10.

35. Rolle 1667, 153–56. Jour. 47, p. 144 (1671-08-23). *London and its environs* 1761, vol. 2, 289. Reddaway 1940, 283.

36. B.R. 1730, 102.

37. NR Transcript minutes Nov. 1619–Sept. 1692, LMA ACC/2558/NR/13/007, folder 3, pp. 163–64 (1656).

38. NR Transcript minutes Nov. 1619–Sept. 1692, LMA ACC/2558/NR/13/007, folder 3, p. 177 (1664). See also LMA ACC/2558/NR/13/054/1–6, cited in R. Ward 2003, 49.

39. NR Letters, LMA ACC/2558/NR/13/037/004–5 (1668-05-21), (1674-02-03), and (1675-06-16).

40. Rep. 49, p. 16 (1669-11-18). *Extracts from the books of the Mayor* 1734, 21–23. Decree of sewers, made at St Andrew, Middx, respecting the navigation of the River Lea to the New River by Chalk Island, NA C 225/2/57 (1669-11-09). Rep. 76, 101 (1670-03-16). JHL vol. 25, pp. 400–410 (1739-06-05).

41. Petty 1683, 32.

42. Gould 1689, 185.

43. NR Book of dividends 1633–1892, LMA ACC/1953/A/258.

44. Rudden 1985, 98.

45. Rudden 1985, 47–48.

46. Report, BL Add MS 70341 (1710?).

47. NR Audit book 1769-72, LMA ACC/2558/NR/09/117.

48. Rudden 1985, 18–19, 58–67.

49. Rudden 1985, 18–19.

50. The total fees are recorded in NA LR 2/27–34. Rudden 1985, 17.

51. Rudden 1985, 148.

52. Rudden 1985, 71.

53. The data on long-term government bond yields are from the Bank of England. www.bankofengland.co.uk/publications/Documents/quarterlybulletin/threecenturies ofdata.xls.

54. Charter in appendix F v of Rudden 1985, 92–96.

55. NA PC 2/59, p. 246/128 (1666-12-21), p. 2667/138 (1666/7-01-16).

56. NA SP 29/188 f.217 (1666/7-01-18). CSPD 1666–67, p. 460 (1666/7-01-18). NA PC 2/59, p. 585 (1667-09-18).

57. NR Transcript minutes Nov. 1619–Sept. 1692, LMA ACC/2558/NR/13/007, folder 3, pp. 179a, 184a, 195a, 196a (1667).

58. CSPD 1664–65, p. 72 (1664-11-16). For a list of the street it supplied, see Weinstein 1991, 38.

59. NA PC 2/61, p. 57 (1668-09-02), p. 222 (1668/9-02-28).

60. BL Add MS 5063, No. 50. *Mary Morris v. Thomas Morris* (1667-04-20). Wilkinson and Shadwell 1819, vol. 1, 11–12.

61. NA SP 29/229 f.205 (1667).

62. CSPD 1687–89, pp. 122–23 (1687-12-05).

63. Defoe 1887, 140.

64. Hoppit 2014, 350–52.

65. Bogart 2011, 1083.

66. Macleod 1986. MacLeod 1988, 150.

67. Dickson 1967. Murphy 2009. Roseveare 1991.

68. Michie 2001, ch. 1. Murphy 2013. Murphy 2009. Scott 1910.

69. Carruthers 1996. Clark 1996. See also Coffman 2013.

70. Hoppit 2011.

71. Sussman and Yafeh 2006.

72. Murphy 2014, 329–31.

73. Murphy 2009, 220.

74. Murphy 2009, ch. 3.

75. Murphy 2009, 31–36. Scott 1910, vol. 1, 347–49.

76. Michie 2001, ch. 1. Murphy 2013. Murphy 2009. Scott 1910.

77. Murphy 2009, ch. 3.

78. Michie 2001, ch. 1. Murphy 2013. Murphy 2009, ch. 1. Scott 1910.

79. Anderson 1764, 292.

80. Bogart 2011, 1081.

81. J. Ward 1974. Albert, Freeman, and Aldcroft 1983.

82. J. Thomas 1979, 110.

83. Macleod 1986, 560–61, 565. J. Thomas 1979.

84. Dickinson 1954, 49. Rep 81, p. 103 (1675/6-02-20). Lysons 1795, vol. 3, 383–90. NA PC 2/65, p. 2 (1675-10-01). J. Thomas 1979, 120–25.

85. RCHM 1894, 32–33. JHC vol. 10, p. 679 (1691/2-02-18). JHL vol. 15, p. 93 (1691/2-02-24). Scott 1910, vol. 3, 32. 3 William and Mary, c. 37.

86. Rep. 80, p. 187 (1675-05-05). JHL vol. 14, p. 618 (1690/1-01-05).

87. NA PC 2/64, p. 410 (1675-05-05).

88. Murray 1883. Dickinson 1954, 47–49. Murphy 2013, 275. JHC vol. 10, 502–3, 508 (1690-12-09 and 12). Clifford 1885, vol.2, 82–83.

89. Hatton 1708, 791. Salmon 1743, vol. 1, 296. Stow and Strype 1720, 66. NA PC 2/64, pp. 280, 289 (1674-09-28). NA PC 2/66, pp. 31, 158 (1677).

90. Kellett 1963.

91. Sheppard 1998, 141.

92. Statutes of the Realm: vol. 6: 1685-94, 464-70. William and Mary, 1694, c. 10: An Act for Relief of the Orphans and other Creditors of the City of London.

93. J. Brown 1779, 134.

94. 35 Henry VIII c. 10. 1543.

95. Alchin MS, LMA COL/AC/06/014/002, #2 pp. 2–3. Rep. 97, p. 258 (1693-04-06). Scott 1910, vol. 3, 4–7. Matthews 1835, 13. Much of its history is recounted in Jour. 71, p. 140 (1757-02-03).

96. Jour. 57, p. 344 (1735-06-18).

97. Rep. 98, p. 156. Stow and Strype 1720, vol. 1, 27.

98. Rep. 100, p. 276 (1697-07-06). Rep. 106, pp. 278, 285 (1702-04-28). Rep. 108, p. 13 (1703-11-11). For the Southwark waters, see Rep. 108, p. 116 (1704–??). Rep. 108, p. 673 (1704-10-26). Rep 111, p. 10 (1706-11-19), p. 39 (1706-12-10).

99. Rep. 103, p. 135 (1698-01-31). Rep. 104, p. 438 (1700-07-16). Park 1814, 74. Bevan 1884, 46.

100. Evan Jones to Robert Harley, NUL Pw2 Hy 336 (1701-10-01), cited at R. Ward 2003, 110.

101. Jour. 71, p. 140 (1757-02-03).

102. Waters to the south were leased to another group, who created the Ravensbourne Waterworks. Dickinson 1954, 52–54.

103. Alchin MS, LMA COL/AC/06/014/011, p. 27. Mayor &c V Richmond et D'Lannoy (Conduit Waters), LMA COL/CCS/SO/01/10/002, p. 15.

104. Rep. 109, p. 52 (1704-12-05), p. 123 (1704/5-02-22), p. 200 (1704/5-03-20). Conduit waters lawsuit, LMA COL/SJ/16/057.

105. Stow and Strype 1720, vol. 2, 27. Rep. 107, p. 407 (1703-06-29). Rep. 108, p. 541 (1704-09-05).

106. *Morris v. Soame*, NA C 5/257/19.

107. NA C 5/257/19, p. 11.

108. Winfield 2007, 2–4.

109. Stow and Strype 1720, vol. 2, 27.

110. Hatton 1708, vol. 2, 791. NA C 5/257/19, p. 3.

111. J. Brown 1779, 131–40. Petition from the Common Council, 1702?, New river and Thames: petition of common council to the queen, LMA COL/SJ/16/026. See also NA PC 2/79, p. 367 (1703-04-24).

112. Rep. 107, p. 425 (1703-07-06), p. 457 (1703-07-20), p. 488 (1703-07-29). Rep. 108, p. 127 (1703/4-02-03), p. 133 (1703/4-02-08). "J. Stafford and others, appellants. Mayor, commonalty, &c. of London, respondents. The Appellants case," 1719, BL shelfmark 19.h.1.(140.). J. Brown 1779, 131–40. R. Ward 2003, ch. 10.

113. Jour. 53, pp. 115, 118, 449–50, 456, 488–92 (1701). NA C 5/257/19, pp. 25, 31–32. Rep. 106, pp. 492, 519, 542 (1702). Rep 107, pp. 5, 32, 35, 63, 96, 103, 195, 196, 244, 247 (1702). Rep. 107, pp. 425, 436, 457, 473, 488, 535, 544, 564, 585 (1703). Morris 1703.

114. RCHM 1897, 51–52.

115. RCHM 1897, 51–52. LBWW lease for first arch of London Bridge, 1581, LMA ACC/2558/MW/C/15/222/2.

116. J. Brown 1779, 131–40. Petition from the Common Council, 1702?, LMA COL/SJ/16/026. See also NA PC 2/79, p. 367 (1703-04-24).

117. Petition from the Common Council 1702?, LMA COL/SJ/16/026.

118. Hatton 1708, vol. 2, 791.

119. Marylebone and Paddington water conduits rentals, 1747–55 and 1756–70, LMA COL/CHD/RN/02/01/026.

120. Carr 1913, cxxvi. JHC vol. 10, 502–3 (1690-12-09 and 12).

121. CSPD 1689–95, p. 106 (1693/4-03-08).

122. CSPD 1694–95, p. 375 (1694/5-01-05).

123. NA PC 2/76, pp. 245, 280, 323 (1695-12-26).

124. London County Council et al. 1937, 21–26.

125. Aubrey and Clark 1898, vol. 1, 255.

126. NR Transcript minutes Nov. 1619–Sept. 1692, LMA ACC/2558/NR/13/007, folder 4, p. 205 (1688-05-17).

127. Rep. 94, p. 65 (1688-11-20).

128. Evan Jones to Robert Harley, NUL Pw2 Hy 336 (1701-10-01), cited at R. Ward 2003, 110.

129. Report of John Lowthorp, BL Add MS 70341, p. 24 (1704–04).

130. BL Add MS 70341, pp. 3, 5 (1710).

131. NR Book of dividends 1633–1892, LMA ACC/1953/A/258.

132. Dickinson 1954, 88. 21 George II c. 8.

133. Scott 1910, vol. 1, 409–21.

134. Harris 2000, ch. 3. Freeman, Pearson, and Taylor 2012, 22–28.

135. NA T 1/248, pp. 126–28 (1724-09-30). van Lieshout 2012, 137, citing NA T/1/248, p. 30.

136. Clifford 1885, vol. 2, 85. JHC vol. 19, p. 284 (1719/20-02-27) and p. 297 (1719/20-03-09).

137. JHC vol. 19, p. 315 (1719/20-03-22), p. 321 (1720-03-28), p. 322 (1720-03-29), p. 324 (1720-03-30), p. 325–26 (1720-03-31).

138. BL Lansdowne 846, fol. 189, 1721?.

139. JHC vol. 19, p. 417 (1720/1-02-07). JHC vol. 20, p. 292 (1723/4-03-12).

140. NA T 1/248, p. 126 (1724-09-30). JHC vol. 20, p. 313 (1724-11-19), p. 372 (1725-01-18). JHC vol. 20, p. 382 (1724/5-01-26), p. 385 (1724/5-01-28), p. 415 (1724/5-02-18), p. 416 (1724/5-02-19), p. 418 (1724/5-02-22).

141. Clifford 1885, vol.2, 87–90. JHC vol. 19, p. 396 (1720/1-01-06), p. 417 (1720/1-02-07).

142. JHC vol. 19, p. 526 (1721-04-24).

143. JHC vol. 19, pp. 580–81 (1721-06-07), p. 584 (1721-06-08), p. 587 (1721-06-12), p. 603 (1721/2-03-20).

144. JHC vol. 19, p. 587 (1721-06-12).

145. JHC vol. 19, pp. 580–81 (1721-06-07).

146. Clifford 1885, vol. 2, 90. See also Stewart 1992, 326–33.

147. JHC vol. 19, p. 717 (1721/2-01-18), p. 727 (1721/2-01-25), p. 739 (1721/2-02-09). An act for better supplying the city and liberty of Westminster and parts adjacent with water. 8 George I c. 28. Sisley 1899, 175–89.

148. Chelsea Waterworks Board Minutes, LMA ACC/2558/CH/01/001, pp. 10, 29, 39 (3rd call), 182. 1725-04-03, 1725-05-27, and 1725/6-01-27. The original price is not stated explicitly, and some calls were made before the minutes began. However, the third one was recorded in April 1725, with one further call.

149. Chelsea Waterworks Board Minutes, LMA ACC/2558/CH/01/004, p. 136 (1730/1-03-10).

150. NA T 1/253, pp. 78, 79 (1725-06-19). Switzer 1729 vol. 2, 323. The reservoir in St. James's Park is now in Green Park according to Dickinson 1954, 55–57.

151. Chelsea Waterworks Board Minutes, LMA ACC/2558/CH/01/002, p. 205 (1727/8-01-04).

152. NA T 1/252, pp. 276, 279 (1725-05-04).

153. Chelsea Waterworks Board Minutes, LMA ACC/2558/CH/01/002, p. 62 (1726-12-17) and p. 199 (1727-12-14). van Lieshout 2012, 193, citing Millbank Waterworks, list of tenants, LMA/ACC/2558/MW/C/15/99, 1715.

154. Chelsea Waterworks Board Minutes, LMA ACC/2558/CH/01/002, p. 62 (1726-12-17). ACC/2558/CH/01/003, p. 66 (1729/30-02-06).

155. Chelsea Waterworks Board Minutes, LMA ACC/2558/CH/01/004, p. 177 (1731-05-29) and p. 196 (1731-08-19). Salmon 1743, 296–97. Stewart 1992, 291–92, 359.

156. Power for increasing the joint stock of the governor and company of Chelsea Waterworks. 7 George II (1733-10-11). Chelsea Waterworks Board Minutes, LMA ACC/2558/CH/01/005, p. 218 (1733-10-31).

157. Chelsea Waterworks Board Minutes, LMA ACC/2558/CH/01/007, p. 56 (1735-10-29), p. 146 (1736-05-5).

158. Chelsea Waterworks Board Minutes, LMA ACC/2558/CH/01/009, p. 170 (1741-12-17). See Report from the select committee on the supply of water to the metropolis, 18 May 1821, PP V (2), 219.

159. Chelsea Waterworks Board Minutes, LMA ACC/2558/CH/01/005, p. 137 (1732/3-02-23). ACC/2558/CH/01/008, p. 122 (1737/8-03-01). van Lieshout 2012, 137, 193, citing Westminster Archives, F2489, Hyde Park Waterworks, list of tenants, 1678.

160. Peppercorne 1840, 65. Report of the commissioners appointed by his majesty to inquire into the state of the supply of water in the metropolis. April 21, 1828. 1828, 5. Minutes of evidence taken before the select committee on the supply of water to the metropolis PP (706) 1821, 245.

161. Chelsea Waterworks Board Minutes, LMA ACC/2558/CH/01/010, p. 210 (1746/7-02-27). ACC/2558/CH/01/011, p. 84 (1753-02-10).

162. Maitland 1756, 1272. Dickinson 1954, 87–88. 21 Geo. II, c. 8. JHL vol. 27, p. 195 (1748-03-25).

Chapter Four · *A New Scale of Network with the New River*

1. Hughes 1983. Lorrain 2005. Radkau 1994. Mayntz and Hughes 1988. Coutard 1999. Summerton 1994.

2. These figures are derived from BL Harley 3604 and assume an average yearly fee of somewhat more than one pound. See chapter 6.

3. Calculated from collector's book LMA ACC/2558/NR/12/001, counting number of lines. Also NR Audit book 1769–72, LMA ACC/2558/NR/09/117, Summary of half year rents pic. 10, shows a net half yearly rental income of £18,898. At twelve shillings per house per six months, this means around 32,500 buildings.

4. See appendix A for more data and sources.

5. Stow and Strype 1720, vol. 1, 28. See also Richards, Payne, and Soper 1899, 10–11. Thumb 1746, 135.

6. Salmon 1743, vol. 1, 295.

7. Voltaire 1784, vol. 60, 241. "Je voudrais que toutes les maisons de Paris eussent de l'eau, comme celles de Londres."

8. "Diese hebt das Wasser aus der Themse welches, nachher durch eine unendliche Menge Röhren durch alle Strassen, und fast in alle Häuser dieses Theils der Stadt geleitet wird. Kaum ist auch vielleicht in ganz Europa ein Ort, der so überflüßig mit Wasser versorgt wäre, als eben London. . . . In jeder Strasse gehen in der Mitte zwey Wasserröhren unter der Erde, von welchen wieder Nebenröhren nach jeden Hause gehen." Fabricius 1784, 191–92.

9. Grimm 1775, vol. 2, 378. Volkmann 1782, vol. 2, 234.

10. Périer 1777, 1–2. See also *Prospectus de la fourniture et distribution des eaux de la Seine, à Paris, par les machines à feu* 1781, and Vachette and Vachette 1791 [1797].

11. Coyer 1783, vol. 5, 111. Letter 1777-04-04.

12. Archenholz 1785, vol. 1, 134.

13. Nugent 1749, 88. See also Dézallier d'Argenville 1747, 350.

14. Auxiron 1765, 27–28.

15. Geels 2005, 371–73, 376–77. The first cities were Amsterdam in 1853 and Den Helder in 1856. They were financed by British investors.

16. Meng 1993. Grewe 2000, 151–60.

17. Colston 1890, 25–26, 42ff. Private supply was a luxury, see p. 45.

18. Report from Evan Jones, 1701-10-01, cited at R. Ward 2003, 118.

19. Proposal, 1704/05, BL Add MS 70341.

20. Proposal, 1705/11, BL Add MS 70341.

21. Ephraim Green report, LMA ACC/2558/NR/14/001 (4) (1705-11-07).

22. See the letter from Lord Clarendon to Robert Harley, RCHM 1894, 609 (1699-10-28); R. Ward 2003, 110.

23. Some considerations and observations on the present posture of affairs in the New River Company, ca. 1710, BL Add MS 70341.

24. Reports and proposals to NR company, LMA ACC/2558/NR/14/001 (06) (1710/1-02-01). Revenues declined from around £13,300 per year between 1704 and 1707 to £11,200 per year from 1707 to 1710.

25. Downes 2004.

26. Hunter 2013, 198–204.

27. Wren 1753. The original date of Wren's report is uncertain, but it was between 1680, when Soho Square was first built, and 1708, when the upper pond was constructed. The date is probably around 1702 because it refers to a proposal to build the upper pond, a suggestion probably first made around 1700. See R. Ward 2003, ch. 12.

28. Wren 1753.

29. Wren 1753.

30. R. Ward 2003, 125. NUL Pw 2 Hy 336.

31. Pumfrey 1995, 145–46. *The Raymond and Beverly Sackler Archive Resource* 2013. Lowthorp 1699. Stewart 1992, 327.

32. Lowthorp reports 1704/05 and 1706/05/18, BL Add MS 70341.

33. Lowthorp report 1704/06/09, BL Add MS 70341. See also NR pre-1769 archives, LMA ACC/2558/NR/13/001 (5) (1709-08-22).

34. Lowthorp report 1705/11, BL Add MS 70341.

35. De Laune and Stow 1681, 210.

36. Lowthorp reports 1706/05/18 and 1711/03/14, BL Add MS 70341. Also proposals in Reports and proposals to NR company, LMA ACC/2558/NR/14/001 (06) (1710/1-02-01).

37. Report, 1711/03/14, BL Add MS 70341.

38. NR Transcript minutes Nov. 1619–Sept. 1692, LMA ACC/2558/NR/13/007, folder 1, p. 12 (1622-06-05). See also p. 14a (1622-11).

39. NR Transcript minutes Nov. 1619–Sept. 1692, LMA ACC/2558/NR/13/007, folder 1, p. 40a, (1632-11-14), p. 42, (1633-05-22), p. 75 (1633-06-24), p. 44 (1633-07-01). Sharpe 1894, vol. 2, pp. 24ff., 1634-10-02.

40. NR Transcript minutes Nov. 1619–Sept. 1692, LMA ACC/2558/NR/13/007, folder 2, p. 131 (1633-05-12), p. 137 (1637).

41. NR Transcript minutes Nov. 1619–Sept. 1692, LMA ACC/2558/NR/13/007, folder 3, p. 123a (1665-04-11), p. 199 (1669).

42. NR Transcript minutes Nov. 1619–Sept. 1692, LMA ACC/2558/NR/13/007, folder 3, p. 184a. CSPD 1667, p. 471 (1667-09-19). Green 1717, 19.

43. Pollard 1965, 102. J. Wilson 1995, 36. Kirby 1994, 126–28.

44. R. Ward 2003, 142. Reports and proposals to NR company, ACC/2558/NR/14/001 #5 (1711-07-16), #6 (1710/1-02-01), #7 (1711-05-31). The term *engineers* was used in the seventeenth century: Morland and Morland 1697, 52.

45. Woodley 2004.

46. Rep. 76, p. 101 (1670-03-16). Report on the River Lea, ACC/2558/MW/C/15/321 (1672-08-30). Rudden 1985, 32.

47. JHL vol. 25, pp. 409–10 (1738-06-05), pp. 412–13 (1739-06-07). *Extracts from the books of the Mayor* 1734.

48. Walton 1738. MB1, p. 20 (1736-03-17), p. 88 (1773-07-22), p. 96 (1774-01-18).

49. Walton 1738. 12 George II c. 32. River Lea act 1739, LMA ACC/2558/MW/C/15/051.

50. MB2, p. 104 (1781-10-4). MB3, p. 242 (1791-12-01). MB4, p. 26 (1792-11-08). MB4, p. 58 (1793-07-25). MB4, p. 219 (1797-06-15).

51. MB3, p. 45 (1787-04-19).

52. By the King. A Proclamation for the Careful Custody and Well Ordering of the New River Brought from Chadwell and Amwell to the North Parts of the City of London 1669. See similar proclamations from 1686, 1689, 1703, 1715.

53. MB6, p. 54 (1805-06-13).

54. MB3, p. 88 (1788-06-26).

55. MB6, p. 54 (1805-06-13). MB3, p. 253 (1792-02-23). MB4, p. 90 (1794-04-03). MB5, p. 158 (1801-09-03). Also manifold ditch: MB4, p. 138 (1794-04-23). MB5, p. 319 (1804-04-26). Neighbor's land: MB6, p. 66 (1805-07-25).

56. MB5, p. 158 (1801-09-03).

57. MB3, p. 110 (1789-01-15).

58. MB4, p. 117 (1795-01-26).

59. MB2, p. 188 (1783-07-17).

60. *A description of the county of Middlesex* 1775, 148.

61. Switzer 1729, vol. 2, 319. Dickinson 1954, 51, citing BL Stowe MS 747 f. 63.

62. George Sorocold proposal, LMA ACC/2558/MW/C/15/199. F. Williamson 1936. F. Williamson and Crump 1945. Fairclough 2004.

63. R. Ward 2003, 130, 172. Dickinson 1954, 38. Agreement with the Earl of Clarendon to let the New River Company construct a pond in the Mantles, LMA ACC/2558/NR/13/314/6 (1708-03-28). For other leases, see LMA ACC/2558/NR/13/314/6; ACC/2558/NR/13/37/1–14; ACC/2558/NR/13/285; ACC/2558/MW/C/15/72.

64. Sorocold report, LMA ACC/2558/MW/C/15/199. Defoe 1725. R. Ward 2003, 132, 137. Chomel 1725, vol. 2, article on "windmil." ACC/2558/NR/13/007, folder 4, p. 266 (1742-10-20).

65. MB2 p. 54 (1780-11-10).

66. NR Transcript minutes Nov. 1619–Sept. 1692, LMA ACC/2558/NR/13/007, folder 4, p. 347. Dickinson 1954, 66, 70.

67. MB4, p. 14 (1792-08-16). MB4, p. 14 (1792-08-23).

68. MB1, p. 97 (1774-02-01). MB3, p. 227 (1791-08-18).

69. MB2, p. 163 (1783-01-09).

70. Chelsea Waterworks Board Minutes, LMA ACC/2558/CH/01/004 (1731-05-29 and 1731-08-19). Stewart 1992, 291, 338–59. Castle 1735, 23–25. Salmon 1743, vol. 2, 296–97. Farey 1827, 117; Dickinson 1954, 59–63.

71. Dickinson 1954, 51, 57.

72. Farey 1827, 245, 253, 255.

73. See Dickinson 1954. All new engines installed after 1780 are listed as steam engines.

74. Dickinson 1954, 80–82.

75. Dickinson 1954, 67–71. Farey 1827.

76. Dickinson 1954, 70–71. Farey 1827, 715–23.

77. Dickinson 1954, 77.

78. Dickinson 1954, 86, 88–89, 99.

79. This is derived from market share data from chapter 6. 1750: CW + YB + SW = 9% + 1% + 5%. 1775: NR(Upper pond) + CW + YB + SW = 3% + 9% + 6% + 5% 1800: NR(Upper pond) + CW + YB + SW + LA = 5% + 9% + 3% + 12% + 5% 1820: All except LBWW and NR(Lower ponds) = 100% – 32% – 7%. The New River supplied about 57,897 tuns per day in 1780 according to Mylne. Boulton & Watt estimated its engine pumped about 3,144 tuns per day in 1788, or about 5 percent. Total market share was 63 percent. The number from 1821 is around 11 percent, based on the evidence given in the PP p. 21. The figure for 1800 was interpolated between the 1775 and 1820 values.

80. Riders in use by 1742: M1, p. 11 (1742-04-07).

81. MB1, p. 26 (1780-03-17).

82. MB2, p. 155 (1782-11-14). MB2, p. 166 (1783-01-28).

83. MB2, p. 168 (1783-02-04). MB2, p. 170 (1783-02-18 and 1783-02-25).

84. MB2, p. 244 (1784-06-09).

85. MB3, p. 15 (1786-07-06). MB3, p. 105 (1788-11-13). MB3, p. 100 (1788-11-20). MB3, p. 121 (1789-04-14). MB5, p. 40 (1799-05-30). MB3, p. 256 (1792-04-05).

86. MB3, p. 55 (1787-08-09).

87. MB3, p. 59 (1787-09-13).

88. MB3, p. 98 (1788-09-18). MB3, p. 66 (1787-11-29). MB4, p. 207 (1797-02-02).

89. MB3, p. 159 (1790-03-18). MB4, p. 136 (1794-07-09). MB4, p. 215 (1797-04-27).

90. Castle 1735. Castle 2012.

91. Smeaton 1812, vol. 1, 7–11.

92. MB4, p. 111 (1794-11-20). MB5, p. 95 (1800-05-08).

93. Du Buat 1786, vol. 1, 141–45, 322–33. Bossut 1786, vol. 2, 130–202.

94. Robison 1797, 898–902.

95. RLW, pp. 54–57 (1745-09-29).

96. MB3, p. 182 (1790-08-19). MB3, p. 191 (1790-10-28).

97. MB4, p. 7 (1792-07-05).

98. MB4, p. 247 (1798-03-08).

99. MB5, p. 191 (1802-07-15).

100. MB2, p. 175 (1783-03-18). MB6, p. 210 (1807-07-09).

101. MB3, p. 256 (1792-04-05).

102. MB3, p. 59 (1787-09-13).

103. MB3, p. 63 (1787-11-08).

104. MB3, p. 36 (1787-02-15).

105. MB3, p. 37 (1787-02-22).

106. MB3, p. 65 (1787-11-157). MB3, p. 83 (1788-06-05). MB3, p. 121 (1789-04-14).

107. MB3, p. 111 (1789-01-22). MB3, p. 112 (1789-02-05). MB3, p. 114 (1789-02-19).

108. MB3, p. 113 (1789-02-19).

109. MB3, p. 171 (1790-04-20).

110. MB4, p. 17 (1792-09-20). MB4, p. 50 (1793-04-16).

111. For one example see MB2, p. 141 (1782-08-08).

112. MB3, p. 87 (1788-06-19).

113. MB3, p. 101 (1788-12-04).

114. MB4, p. 106 (1794-10-09).

115. MB3, p. 11 (1786-06-01).

116. MB3, p. 55 (1787-08-02).

117. MB3, p. 21 (1786-08-04).

118. MB3, p. 38 (1787-03-01).

119. MB5, p. 43 (1799-07-04).

120. MB1, p. 3 (1675).

121. MB1, p. 11 (1742-04-07).

122. MB2, p. 9 (1778-12-17). MB1, p. 34 (1770-03-27).

123. MB4, p. 30 (1792-11-29).

124. NR Transcript minutes Nov. 1619–Sept. 1692, LMA ACC/2558/NR/13/007, folder 1, p. 15 (1620-11). NR Miscellaneous papers, LMA ACC/2558/NR/05/077, pp. 374–85 (1791). MB6, p. 7 (1805-01-10). MB6, p. 12 (1805-01-24).

125. Wren 1753, 115.

126. NR Transcript minutes Nov. 1619–Sept. 1692, LMA ACC/2558/NR/13/007, folder 3, p. 198 (1668). MB1, p. 1. R. Ward 2003, 95.

127. NR Transcript minutes Nov. 1619–Sept. 1692, LMA ACC/2558/NR/13/007, folder 2, p. 138a (1638). *By the King, a proclamation, for the careful custody and well-ordering of the new river, brought from Chadwell and Amwell to the north parts of the City of London* 1669. *By the King, a proclamation, for the careful custody and well-ordering of the new river, brought from Chadwell and Amwell* 1686. *By the Queen, a proclamation, for the careful custody and well ordering of the new river brought from Chadwell and Amwell* 1703. Petition from the New River, PC 1/2/264 1714-10-27. *By the King, a proclamation, for the careful custody and well-ordering of the new river, brought from Chadwell and Amwell* 1715.

128. MB2, p. 236 (1785-03-17). MB2, p. 240 (1785-04-28). MB2, p. 242 (1785-05-12).

129. MB2, p. 210 (1784-03-25).

130. MB1, p. 129 (1776-02-15).

131. MB4, p. 30 (1792-11-29).

132. MB2, p. 223 (1784-09-23).

133. MB2, p. 252 (1785-08-04).

134. MB3, p. 33 (1787-02-01).

135. MB4, p. 255 (1798-05-10).

136. MB5, p. 110 (1800-08-28).

137. MB6, p. 256 (1808-03-24), MB6, p. 268 (1808-05-19).

138. MB1, p. 2. Repeated in later years, MB2, p. 96 (1781-07-26). MB3, p. 67 (1787-12-13).

139. MB3, p. 8 (1786-04-11).

140. MB3, p. 67 (1787-12-13).

141. LMA ACC/2558/MW/C/15/173 (1594).

142. MB5, p. 59 (1799-09-12).

143. MB2, p. 160 (1782-12-03).

144. MB5, p. 113 (1800-09-11).

145. MB5, p. 29 (1799-04-11).

146. MB3, p. 152 (1790-02-04).

147. MB2, p. 241 (1784-04-12).

148. MB5, p. 178 (1802-02-25). MB5, p. 205 (1802-04-15).

149. MB4, p. 249 (1798-03-22).

150. MB3, p. 152 (1790-02-04).

151. MB3, p. 96 (1788-09-11). MB3, p. 101 (1788-10-16).

152. MB4, p. 250 (1798-04-05).

153. Hatton 1708, 793. Maitland 1739, 630. "Sir Hugh Middleton's Scheme for supplying the City of London with good and wholesome Water" 1751, 312. *London and its environs* 1761, vol. 5, 42. *The Times*, 1791-03-31.

154. MB1, p. 3 (1757).

155. MB1, p. 2 (1716). MB1, p. 156 (1778-04-16).

156. Maitland 1739, 630.

157. NR Letter book, LMA ACC/2558/NR/03/001, p. 309 (1715/6-03-03). MB4, p. 202 (1796-12-15).

158. Chelsea Waterworks Board Minutes, LMA ACC/2558/CH/01/005, p. 179 (1733-06-21).

159. MB4, p. 151 (1794-11-03).

160. MB5, p. 190 (1802-07-08).

161. MB5, p. 241 (1803-02-24).

162. MB5, p. 110 (1800-08-28).

163. MB5, p. 105 (1800-07-24).

164. MB1, p. 156 (1778-04-16).

165. MB1, p. 113 (1774-01-24). MB3, p. 98 (1788-09-11). MB3, p. 117 (1789-03-26).

166. Chelsea Waterworks Board Minutes, LMA ACC/2558/CH/01/012, p. 94 (1764-12-20).

167. MB5, p. 46 (1799-07-11).

168. MB5, p. 142 (1801-04-23). MB5, p. 350 (1804-10-18). MB5, p. 352 (1804-10-25).

169. Report to New River board 1814, LMA ACC/2558/NR/14/002, p. 2 (1814-03).

170. Matthews 1835, 66.

171. NR Report on the state of the works, LMA ACC/2558/NR/14/002 (1814-03).

172. R. Ward 2003, 51.

173. NR Report on the state of the works, LMA ACC/2558/NR/14/002 (1814-03).

174. Corroborated by figures reported in LMA ACC/2558/NR/14/002, p. 20 (1814-03).

175. Clark 2004, 90–105.

176. MB4, p. 98 (1794-07-03). See also G. Jacob 1717, 37.

177. MB5, pp. 49–51 (1799-07-18 and 1799-08-01). *Caledonian Mercury*, 1805-04-25, 1808-10-31, *Aberdeen Journal*, 1805-05-01, 1808-10-26. *Hampshire Chronicle*, 1805-04-01. *Reading Mercury*, 1799-11-11, 1801-05-25.

178. King 2005.

179. Matthews 1835, 70.

180. Gough 1964, 61.

181. MB2, p. 264 (1784-12-08). MB3, p. 10 (1786-04-25). MB3, p. 11 (1786-06-01). MB3, p. 14 (1786-06-22).

182. MB3, p. 23 (1786-09-28).

183. MB3, p. 25 (1786-10-19). MB3, p. 55 (1787-08-02).

184. MB3, p. 30 (1786-12-14).

185. MB4, p. 99 (1794-07-17). Martin 1813, 480.

186. Switzer 1729, vol. 1, 118.

187. "Sir Hugh Middleton's Scheme for supplying the City of London with good and wholesome Water" 1751. Maitland 1739, 630. MB5, p. 32 (1799-04-23).

188. MB2, p. 115 (1774-02-09). MB5, p. 84 (1800-02-13).

189. MB1 p. 47 (1771-02-21).

190. MB5, p. 152 (1801-07-16). MB5, p. 173 (1802-01-14).

191. MB5, p. 294 (1803-11-17).

192. *Middlesex Journal or Chronicle of Liberty*, 1770-07-12.

193. MB3, p. 182 (1790-08-19). MB4, p. 20 (1792-10-18).

194. MB4, p. 120 (1795-03-05).

195. MB1, p. 129 (1776-02-15). MB3, p. 63 (1787-11-08).

196. MB4, p. 261 (1798-07-19).

197. Among many others, see MB3, p. 126 (1789-07-09). MB5, p. 64 (1799-10-10).

198. MB2, p. 141 (1782-08-01).

199. MB5, p. 100 (1800-06-19).

200. MB5, p. 34 (1799-05-02).

201. MB4, p. 202 (1796-12-15).

202. MB2, p. 139 (1782-07-04).

203. MB3, p. 10 (1786-04-25). MB5, p. 95 (1800-05-08).

204. MB4, p. 79 (1794-01-16).

205. MB4, p. 187 (1796-08-04). MB5, p. 90 (1800-03-27).

206. MB4, p. 246 (1798-02-08).

207. MB3, p. 42 (1787-03-29).

208. MB3, p. 62 (1787-10-25).

209. MB5, p. 143 (1801-04-30).

210. MB3, p. 55 (1787-08-02).

211. *The builder's dictionary or, Gentleman and Architect's companion* 1734, 461.

212. *A new and complete dictionary of arts and sciences* 1763, vol. 3, 2486.

213. Desaguliers 1726.

214. MB1, p. 158 (1782-12-03).

215. MB4, p. 88 (1794-03-13). MB4, p. 256 (1798-05-17).

216. MB3, p. 21 (1786-08-17).

217. MB5, p. 358 (1804-11-15).

218. MB5, p. 220 (1802-01-13).

219. MB2, p. 77 (1781-05-03). MB2, p. 79 (1781-05-10).

220. *The London Chronicle, or, Universal Evening Post*, p. 109, 1763-01-29. Trusler 1786, 26.

221. MB3, p. 109 (1788-01-08).

222. MB2, p. 205 (1784-02-05).

223. MB5, p. 29 (1799-04-11).

224. MB4, p. 117 (1795-01-26). MB4, p. 117 (1795-01-29). MB4, p. 118 (1794-02-05). MB4, p. 115 (1795-02-19). MB4, p. 119 (1795-02-26). MB4, p. 125 (1795-04-16). MB4, p. 128 (1795-04-07).

225. MB3 p. 38 (1787-03-01). "Sir Hugh Middleton's Scheme for supplying the City of London with good and wholesome Water" 1751.

226. MB1, p. 98 (1774-03-15).

227. MB1, pp. 2–4.

228. MB4, p. 232 (1797-09-07).

229. MB4, p. 232 (1797-09-07).

230. For this point see also MB1, p. 3 (1703).

231. MB2, pp. 157–60 (1782-12-03).

232. MB4, p. 132 (1795-06-11). MB4, p. 151 (1795-11-03). MB4, p. 157 (1795-12-10).

233. MB3, p. 6 (1786-04-27). MB4, p. 79 (1794-01-16).

234. MB1, p. 114 (1774-02-07).

235. MB3, p. 134 (1789-09-17).

236. MB1, p. 12 (1752-07-29). MB1, p. 12 (1758-03-03).

237. Rudden 1985, 101–7.

238. NR Miscellaneous papers, LMA ACC/2558/NR/05/066, p. 33.

239. Vernon 1726, vol. 2, 431–32. Rudden 1985, 104.

240. NR Miscellaneous papers, LMA ACC/2558/NR/05/066, p. 33.

Chapter Five · The London Bridge Waterworks and Other Companies in the Eighteenth Century

1. R. Ward 2003, 118. Hatton 1708, vol. 2, 791–93.

2. LBWW Samuel Hearne, report on the state of the company, LMA ACC/2558/MW/C/15/102, pp. 10, 43.

3. LBWW Committee of managers minute book, LMA ACC/2558/LB/01/002, p. 15 (1779/12/10). Wicksteed 1835, 7.

4. See chapter 6.

5. Minutes of evidence taken before the select committee on the supply of water to the metropolis PP (706) 1821, 246.

6. Tynan 2002, 348.

7. Wicksteed 1835, 11. Sisley 1899, 9.

8. Mylne's notebook, LMA ACC/2558/NR/13/188, p. 19. The figures seem to date to 1767. Adolphus 1818, vol. 3, 537.

9. John Rocque, *London, Westminster and Southwark*, 1746.

10. Map entitled *A Plan of the Cities of London and Westminster, and Borough of Southwark, With the New Buildings* 1767.

11. Stewart 1992, 328, 352. Saussure and Van Muyden 1902, pp. 138–39, 156–58 (1726-06-14 and 1726-10-29).

12. *Daily Journal*, 1733-07-11 and 12. Salmon 1743, 296–97.

13. Map entitled *A Plan of the Cities of London and Westminster, and Borough of Southwark, With the New Buildings* 1767.

14. Dickinson 1954, 79–83. van Lieshout 2012, 74–75. Concanen and Morgan 1795, 233–34.

15. Dickinson 1954, 83. 25 Geo. III, c. 89.

16. van Lieshout 2012, 82.

17. See chapter 6.

18. Reddaway 1938, 217–18.

19. Jenkins 1895, 261–62. Bate 1635, 57–60. BL Add. MS. 5063, No. 50. *Mary Morris v. Thomas Morris. London and its environs* 1761, vol. 4, 87.

20. Wilkinson and Shadwell 1819, vol.1, 11–12.

21. Hatton 1708, vol. 2, 791–93.

22. Hatton 1708, vol. 2, 791–93.

23. ACC/2558/NR/14/001 #06 (1710/1-02-01).

24. Hatton 1708, vol. 2, 791. Beighton 1731. F. Williamson 1936.

25. F. Williamson 1936, 65, 82–85. Hatton 1708, vol. 2, 791–92.

26. ACC/2558/NR/14/001 #06 (1710/1-02-01). See also R. Ward 2003, 118–19.

27. Jour. 57, p. 233 (1731-06-29). Thomson 1827, 556.

28. LBWW NR agreement 1738, LMA ACC/2558/MW/C/15/291/01 (1738-06-23).

29. Castle 1735, 25.

30. RLW.

31. RLW, p. 2 (1745-09-29).

32. RLW, p. 4. Saussure and Van Muyden 1902, 83.

33. Bate 1635, 59.

34. Grewe 2000.

35. RLW, pp. 4–5. Beighton 1731.

36. T. Reynolds 1983, 287.

37. King 2005, 20–24.

38. T. Reynolds 1983, 286–320.

39. Water 1819.

40. Beighton 1731. See also Dickinson 1954, 26–28.

41. Beighton 1731, 8. He did not include a mention of the wheels in the second arch.

42. RLW, pp. 15–22.

43. RLW, pp. 19–22.

44. RLW, pp. 23–28.

45. RLW, pp. 29–31.

46. RLW, p. 33. Desaguliers gave a much higher figure. Desaguliers 1744, vol. 2, 530.

47. RLW, p. 10.

48. The New River Company at this time supplied about 15 percent to large customers, based on their ledger book from 1769. See chapter 6.

49. RLW, pp. 7–15.

50. RLW, pp. 35–39.

51. RLW, pp. 38–39.

52. RLW, pp. 45–46.

53. RLW, p. 51.

54. RLW, p. 48.

55. RLW, p. 48.

56. RLW, pp. 42–49.

57. Hall 1974. Wengenroth 2003.

58. Bate 1635, 59. Switzer 1729, vol. 2, 319.

59. Switzer 1729, vol. 2, 319.

60. RLW, pp. 49–50.

61. RLW, p. 51.

62. RLW, p. 50–53.

63. RLW, pp. 53–55.

64. RLW, pp. 55–57.

65. RLW, pp. 54–57.

66. RLW, p. 58.

67. RLW, pp. 57–59.

68. RLW, p. 58.

69. BL Add MS 70341, p. 5.

70. NR Case regarding J. Whishaw 1708–44, LMA ACC/2558/NR/13/041, p. 23.

71. RLW, p. 121. Matthews 1835, 66.

72. *Penny Cyclopaedia of the Society for the Diffusion of Useful Knowledge* 1833, vol. 5, 416.

73. Willan 1936, 67.

74. RLW, pp. 122–25.

75. RLW, pp. 43, 122–25.

76. RLW, pp. 122–25.

77. Text of lease in LMA ACC/2558/MW/C/15/222/2.

78. Smeaton 1812, vol. 2, pp. 29–30. LBWW Committee of managers minute book, LMA ACC/2558/LB/01/002, p. 3 (1779/11/03).

79. Water 1819.

80. Thomson 1827, 556. Jour. 63, pp. 369–70 (1766-11-28).

81. Jour. 63, p. 362 (1766-11-28). Jour. 64, p. 86 (1766-12-16). Thomson 1827, 555–60.

82. Jour. 64, pp. 119–24 (1767-02-25).

83. Jour. 64, pp. 128–29 (1767-03-13).

84. Jour. 64, pp. 152–60 (1767-06-23).

85. Moher 1986. Smeaton 1812, vol. 2, 27–28. Water 1819.

86. Smeaton 1812, vol. 2, 22ff. Water 1819. Dickinson 1954, 28–29. LBWW Committee of managers minute book 4, LMA ACC/2558/LB/01/004 (1788-11-14 and 1789-11-13).

87. NR Letter book, LMA ACC/2558/NR/03/001, p. 19 (1779-11-04).

88. LBWW Committee of managers minute book 2, LMA ACC/2558/LB/01/002 (1779-11-03 to 1779-12-17). Minute book 3, LMA ACC/2558/LB/01/003 (1780-11-03 to 1780-11-10, and 1782-07-19).

89. Moher 1986. Quote on p. 66. Chrimes 2002a.

90. LMA ACC/2558/LB/01/002 (1779-12-10).

91. LBWW Minute book 3, LMA ACC/2558/LB/01/003 (1784-03-26, 1784-05-21, and 1786-06-30).

92. Moher 1986, 65–66. LBWW Minute book 4, LMA ACC/2558/LB/01/004 (1792-04-27).

93. LBWW Minute book 4, LMA ACC/2558/LB/01/004 (1794-11-14).

94. Schwarz 2000, 648.

95. Krünitz 1801, 331.

96. LBWW Minute book 3, LMA ACC/2558/LB/01/003 (1784-10-22 and 1784-11-26). Minute book 4, LMA ACC/2558/LB/01/004 (1794-11-14).

97. Scott 1910, vol. 3, 16. LBWW Minute book 4, LMA ACC/2558/LB/01/004 (1794-11-14). Minute book 5, LMA ACC/2558/LB/01/005 (1798-11-16).

98. Assignment by West Ham proprietors to Shadwell Waterworks of all mains, pipes etc., ACC/3077/006 (1792-08-23).

Chapter 6 · Consumption

1. Saussure and Van Muyden 1902, 155–56, 157–58 (1726-10-29).
2. Landers 1993, 57–58.
3. Landers 1993, 53-55. Maitland 1739, 519–32. Richard Price 1780, 1-2. Marshall 1832, 3. Hooke and Derham 1726, 168.
4. Richard Price 1780, 1-2. Richard Price and Morgan 1803, vol. 2, 20. A few observations are important in interpreting these figures. Price mentions that landlords sometimes listed a single house even if they owned a number. Price's house counts also include areas south of the Thames not served by the companies listed here; the number there was, however, much smaller than north of the river. The bills of mortality and the window tax count houses and exclude other kinds of buildings. See also Chapman 1994, 47. There is also a figure for the City from 1741: 21,649. Smart 1742, 3.
5. *Universal Chronicle or Weekly Gazette for the Year* 1758, vol. 1, p. 14.
6. Hatton 1708, 791–92. Trusler 1786, 24–25. Weale 1820, 69–71.
7. MB2, p. 172 (1783-02-27).
8. See also Goldsmith and Carter 2015.
9. Williamson 1982. The company's own laborers were paid two shillings per week. MB2, p. 160 (1782-12-03).
10. Chartres 1986, 171.
11. Earle 1989, 269–70.
12. Castle 1735, 14. See chapter 5 on Hearne's volume calculations. Walton 1738, 2.
13. Lowthorp supposed that a house used half a tun per day when looking at raw water flow into the system. Lowthorp report 1704/06/09, BL Add MS 70341. See also NR pre-1769 archives, LMA ACC/2558/NR/13/001 #5 (1709-08-22). Lowthorp also stated that a family uses 2–3 barrels (1/4–3/8 tun) per day when looking at raw water flow into the system. Lowthorp report 1705/11, BL Add MS 70341. Sorocold estimated that net system water per household was half a tun per day. Sorocold proposal dated 1700, LMA ACC/2558/MW/C/15/199. Houses assumed to use one hogshead per day when looking at raw water flow into system. Chelsea Waterworks Board Minutes, LMA ACC/2558/CH/01/009, p. 180 (1741/2-01-28).
14. See appendix A.
15. Environment Canada Report (2013), www.ec.gc.ca/eau-water/default.asp?lang=En&n=F25C70EC-1.
16. Saussure and Van Muyden 1902, 155–56, 157–58 (1726-10-29).
17. Minutes of evidence taken before the select committee on the supply of water to the metropolis (706) 1821, 246.
18. Tabberner 1847, 5.
19. Hassan 1998, 14.
20. Maxwell 2003, 8–10.
21. Girard 1812, 15–16, 22. For the ownership of these and other waters, see p. 101. Beaumont-Maillet 1991, 71–72.
22. Girard 1812, 21, 23. Beaumont-Maillet 1991, 81–83.

23. Girard 1812, 25–27. Beaumont-Maillet 1991, 86–87.

24. Girard 1812, 48, 50.

25. Girard 1812, 42–43. Carbonnier 2006, 438–39. Beaumont-Maillet 1991, 89–94. The city government controlled this pump and some water drawn from the north, while the Crown controlled the earlier bridge pump and the Arceuil aqueduct. See p. 146.

26. Girard 1812, 46, 49.

27. Massounie 2009, 21.

28. Girard 1812, 64.

29. Girard 1812, 70–71.

30. Mirabeau 1786, 72. See also p. 77. The total capacity was estimated at three thousand customers. Girard 1812, 89. Vachette and Vachette 1791 [1797], 6.

31. Chatzis and Coutard 2005, 3. Carbonnier 2006, 440. Girard 1812, 73–89. Beaumont-Maillet 1991, 98–107. Bouchary 1946.

32. Bocquet, Chatzis, and Sander 2008.

33. Hamlin 2000, 726.

34. De Vries 2008, 32–37.

35. De Vries 2008, 52.

36. De Vries 2008, 54–56, and ch. 4.

37. De Vries 2008, ch. 3.

38. Hoppit 2000, ch. 10. Slack 2014. Earle 1989, ch. 10. Horrell 2014, 244–45.

39. Peck 2005. Slack 2014, 149.

40. Slack 2014, 142–47.

41. Wrigley 1985, 688.

42. Broadberry et al. 2014, 194–95.

43. Broadberry et al. 2014, 262. R. Allen 2009, 33–35.

44. Broadberry et al. 2014, 204–5. There are more estimates of real GDP per capita, but others also show increase after 1650. See pp. 250–52. Slack 2014, 10–12. Horrell 2014, 243. See also Borsay 2001 for a discussion of London's special place in the country.

45. Horrell 2014, 247–53. N. McKendrick, Brewer, and Plumb 1982. N. McKendrick 1974.

46. Horrell 2014, 248. Broadberry et al. 2014, 261. Voth 2000.

47. De Vries 2008, 65–70.

48. MB4, p. 247 (1798-02-22).

49. Chelsea Waterworks Board Minutes, LMA ACC/2558/CH/01/005, p. 14 (1731-11-10).

50. Richmond and Turton 1990, 112–13.

51. Richmond and Turton 1990, 336–37.

52. Richmond and Turton 1990, 366–67.

53. Richmond and Turton 1990, 103–4.

54. Mathias 1959, 25.

55. Richmond and Turton 1990, 54–55.

56. Schwarz 1992, 66–67.

57. Mathias 1959, ch. 1.

58. Minutes of evidence taken before the select committee on the supply of water to the metropolis PP (706) 1821, 231.

59. BL Add MS 70341, p. 4 (1710). NR Reports and proposals 1705–50, LMA ACC/2558/NR/14/001 #6 (1710/1-02-01). A tenant on St. John Lane was supplied from 10 PM to 5 AM. MB4, p. 161 (1796-02-21).

60. MB3, p. 96 (1788-09-11). MB3, p. 101 (1788-10-16). MB3 p. 152 (1790-02-04). MB4 p. 250 (1798-04-05).

61. Sheppard 1998, 177. McKellar 1999, ch. 9. Summerson and Colvin 2003, ch. 3.

62. Keene 2001, 32–37. McKellar 1999, ch. 9. Summerson and Colvin 2003, ch. 3.

63. MB4, p. 163 (1796-02-11).

64. Increase supply: MB3, p. 111 (1789-02-22). MB3, p. 112 (1789-02-05). MB3, p. 114 (1789-02-19). Decide on new iron main: MB3, p. 171 (1790-04-20). Extra hours: MB3, p. 37 (1787-02-22). MB3, p. 251 (1792-02-09). MB5, p. 175 (1802-02-04). MB5, p. 79 (1800-01-23). MB5, p. 130 (1801-01-22).

65. Keene 2001a, 35.

66. Keene 2001a, 17.

67. MB1, p. 89 (1773-08-19).

68. MB6, p. 34 (1805-04-04). MB6, p. 44 (1805-05-02). MB6, p. 50 (1805-05-23). MB6, p. 57 (1805-06-27). MB6, p. 84 (1805-11-07). MB6, p. 314 (1809-03-30).

69. NR Reports and proposals 1705–50, LMA ACC/2558/NR/14/001 #6 (1710/1-02-01).

70. Remem. I. 449, p. 553 (1582-12).

71. NR Lease with Thomas Parradine 1604, LMA ACC/2558/MW/SU/01/040/2050. R. Ward 2003, 19.

72. City records, letter book DD, f. 223, cited in Gough 1964, 44.

73. Remem. IV. 46, pp. 556–57 (1616-12-23).

74. JHC vol. 1, p. 745 (1623/4-02-22).

75. Rep. 38, p. 12 (1623-11-03).

76. Gough 1964, 78.

77. Rep. 45, p. 544 (1631-10-20).

78. Jour. 36, p. 36 (1632-12-14), p. 292 (1634-10-02). Sharpe 1894, 26–27.

79. Jour. 39, p. 62 (1639-03-09).

80. Gosling 1643. For a description of the fireplugs, see Fabricius 1784, 192.

81. Blackstone 1957, 26–27.

82. NR Transcript minutes Nov. 1619–Sept. 1692, LMA ACC/2558/NR/13/007, folder 3, p. 159 (1648), p. 164 (1656).

83. Ley 2000, 1–3. Earle 1989, 207. Reddaway 1940. Blackstone 1957, ch. 4.

84. Evans 1987. See also Pearson 2004, 18–21. Blackstone 1957, ch. 5. Relton 1893.

85. Blackstone 1957, 50–61.

86. Entick 1766, vol. 2, 354–55. Blackstone 1957, 61. 6 Anne c. 31.

87. NR Transcript minutes Nov. 1619–Sept. 1692, LMA ACC/2558/NR/13/007, folder 5, pp. 290–91 (1720).

88. Defoe 1725, vol. 2, 148–49.

89. Archenholz 1785, vol. 1, 134. See also Coyer 1783, vol. 5, 111, letter 1777-04-04.

90. Herring 1625, 7.

91. Saussure and Van Muyden 1902, 68.

92. *Lloyd's Evening Post*, 1800-08-08.

93. NR Letter book, LMA ACC/2558/NR/03/001, p. 39 (1784-12-28). Chelsea Waterworks Board Minutes, LMA ACC/2558/CH/01/005, p. 179 (1733-06-21). *Daily Journal*, Issue 2492, 1728/9-01-02.

Chapter 7 · Purity, 1700–1810

1. Melosi 2000. Thorsheim 2006. Hamlin 1998. Worboys 2007. M. Allen 2008. On the historiography, see Hamlin 2012.

2. Cook 2001. Hamlin 1990; 1992; 1994. Taylor and Trentmann 2011. Mukhopadhyay 1981. Goddard and Sheail 2004.

3. Halliday 1999.

4. Jenner 2007.

5. Mullett 1946. Coley 1979; 1982; 1990. Hardy 1984.

6. Hamlin 2000.

7. Brown 2008.

8. Culpeper 1652, 6. W. Roberts 1641a. See also W. Heberden 1768, 89. And more broadly Cockayne 2007.

9. Hardy 1984, 253. Hamlin 1990, 77–79. For earlier notions of water quality, see Hamlin 2000. See also Percival 1769, 65ff. Hutton 1795, 672. Middleton 1798, 5. Haigh 1779, 17.

10. *Extracts from the books of the Mayor* 1734, 13–17.

11. For a detailed description of the legal status of waterways in England, see Willan 1936, 16–51. See also Getzler 2004.

12. Rudden 1985, 288.

13. *By the King, a proclamation, for the careful custody and well-ordering of the new river, brought from Chadwell and Amwell* 1686.

14. NR Transcript minutes Nov. 1619–Sept. 1692, LMA ACC/2558/NR/13/007, p. 205a (1688-04-26). For another incident with a tanner dumping, see also folder 4, p. 242 (1761-08-03).

15. Hatton 1708, 793, reports 12. *London and its environs* 1761, vol. 2, 42, reports 14. Still 14 in 1790, *The Universal British directory of trade and commerce* 1790, 441. 19 in 1805. MB6, p. 78 (1805-10-10). 15 in 1807 MB6, p. 184 (1807-03-05).

16. Maitland 1739, 630.

17. For a list of duties from around 1710, see BL Add MS 70341.

18. E.g., MB2, p. 249 (1784-07-21). See also proposal for increasing the current of the New River, LMA ACC/2558/MW/C/15/320.

19. MB2, p. 95 (1781-04-28). MB3, p. 88 (1788-06-26).

20. MB1, p. 8 (1769-07-27). MB2, p. 33 (1780-04-25). MB2, p. 92 (1781-06-29).

21. MB6, p. 209 (1807-07-09).

22. Owen 1754, 196. *Public Advertiser*, 1774-07-22.

23. Tryon 1683, 112.

24. Harvey 1701, 26.

25. Hancocke 1723. Price 1981, 271. Jenner 1996.

26. Culpeper 1652, 6.

27. Ellis 1734, 21. Ellis 1736, 59.

28. Baker 1743, 75. Turner 2004.

29. MB1, p. 11 (1744-07-18). NR Transcript minutes Nov. 1619–Sept. 1692, LMA ACC/2558/NR/13/007, folder 4, p. 258 (1744-07-18). A similar incident occurred in 1688:

ACC/2558/NR/13/007, folder 1, p. 205a (1688-04-26). And ACC/2558/NR/13/007, folder 4, p. 242 (1761).

30. On cattle plague, or rinderpest, see Spinage 2003, ch. 7.
31. MB1, p. 7 (1769-03-09 and 1769-03-23).
32. MB2, p. 32 (1780-05-25). Letters to William Marshall in NR Letter book [1770–87], LMA ACC/2558/NR/03/001, p. 20 (1780-06-08) and p. 28 (1781-07-12). MB2, p. 36 (1780-06-22). MB2, p. 102 (1781-08-30).
33. MB2, p. 80 (1781-05-10). LMA ACC/2558/NR/13/007, folder 4, p. 253 (1743-09-07).
34. Pringle 1752. Hardy 1984, 256.
35. W. Heberden 1768, 88. E. Heberden 1986. E. Heberden 2004.
36. Porter 1995, 431. J. Wilson 1995, 2.
37. White 1773, 155. Butler 2004.
38. Fothergill and Lettsom 1784, vol. 3, pp. cxvi–cxviii.
39. Letter to Boyer Esq. in NR Letter book [1770–87], LMA ACC/2558/NR/03/001, p. 28 (1781-07-12). Letters to/from Col. Armstrong in NR Letter book [1788–1807], LMA ACC/2558/NR/03/002, pp. 9–11 (1788-12-04, 1788-12-11, 1788-12-18).
40. MB2, p. 131 (1782-04-25). Letter to Bell in NR Letter book [1770–87], LMA ACC/2558/NR/03/001, p. 29 (1782-04-26); letter to John Dawes p. 29 (1782-05-03).
41. MB2, p. 134 (1782-04-30).
42. MB3, p. 36 (1787-02-22).
43. MB3, p. 225 (1791-08-04). Letter to Dubois in NR Letter book [1788–1807], LMA ACC/2558/NR/03/002, p. 20 (1791-08-11).
44. MB5, p. 26 (1799-03-07). Aguilar 1906.
45. MB5, p. 30 (1799-04-11).
46. MB5, p. 52 (1799-08-01).
47. MB6, p. 31 (1805-03-28).
48. MB4, p. 43 (1793-03-28). The same happened in 1806: Letter to Smith in NR Letter book [1788–1807], LMA ACC/2558/NR/03/002, p. 36 (1806-08-21).
49. MB5, p. 186 (1802-05-17). MB5, p. 267 (1803-06-09). MB6, p. 228 (1807-10-22).
50. MB5, p. 186 (1802-05-17).
51. MB5, p. 197 (1802-08-19).
52. MB6, p. 209 (1807-07-09).
53. MB6, p. 344 (1809-10-26). MB6, p. 352 (1809-11-30).
54. MB6, p. 354 (1809-12-14). MB6, pp. 356–57 (1809-12-21).
55. Douglas 1966.
56. Hembry 1990, 161–67, 169, 174. Mullett 1946, 19–25. Jenner 1998.
57. Vigarello and Birrell 1988, 42–44. Smith 2007, 233–36, 248.
58. Thomas 1997, 27. Elias 1994.
59. Elias 1994, 114, 134. Bonneville 1998, 40–41.
60. Gatrell 2007, 457.
61. M. Roberts 1988, 279–81.
62. National archives, LR/2/27–33 (1614-06 and 1616-03-30). R. Ward 2003, 96.
63. *By the King A proclamation for the careful custody and well ordering of the New River brought from Chadwell and Amwell to the north parts of the City of London*, 1686-04-08. Ibid. 1689-09-20.

64. R. Ward 2003, 96.

65. *Public Advertiser*, 1755-06-27.

66. *Gazetteer and New Daily Advertiser*, 1765-08-26.

67. Ibid., 1765-09-04.

68. Ibid., 1766-06-25.

69. Ibid., 1766-07-19.

70. Ibid., 1766-07-23 and 1766-09-10.

71. *London Evening Post*, 1766-09-18.

72. *Public Advertiser*, 1767-07-28.

73. *Gazetteer and New Daily Advertiser*, 1768-05-30 and 1770-08-21. *Public Advertiser*, 1770-08-04. MB1, p. 41, 1770-08-16. The directors got complaints about people washing in the river. MB1, pp. 51–52, 1771-06-06 and 1771-06-13.

74. *Daily Advertiser*, 1772-06-27. *General Evening Post*, 1772-08-13. *Middlesex Journal or Universal Evening Post*, 1773-08-19. *Morning Chronicle and London Advertiser*, 1773-06-16. Assault in mentioned in *Morning Post and Daily Advertiser*, 1776-06-13. *Public Advertiser*, 1776-06-15. Harassment in *Gazetteer and New Daily Advertiser*, 1777-09-27.

75. *Morning Chronicle and London Advertiser*, 1778-07-06. *Public Advertiser*, 1778-07-04.

76. *General Advertiser and Morning Intelligencer*, 1778-08-14.

77. *Morning Chronicle and London Advertiser*, 1778-07-13.

78. Ibid., 1778-07-09.

79. *Public Advertiser*, 1778-07-21.

80. *General Advertiser and Morning Intelligencer*, 1778-08-15. For another response with similar arguments, see ibid., 1778-08-18. For a rebuttal to these, see ibid., 1778-08-20. For another letter arguing that it was specifically the poor being deprived of health because the New River was their only way of washing, see ibid., 1779-08-31.

81. Ibid., 1778-08-24.

82. *Gazetteer and New Daily Advertiser*, 1778-08-22. *General Advertiser and Morning Intelligencer*, 1778-09-14; 1779-06-12; 1779-06-15. *Gazetteer and New Daily Advertiser*, 1779-08-13; 1779-08-20; 1779-08-09; 1779-08-26; 1779-07-27. *Morning Chronicle and London Advertiser*, 1779-07-02; 1779-07-10. *Public Advertiser*, 1779-07-19; 1779-06-26; 1779-08-12.

83. *Gazetteer and New Daily Advertiser*, 1779-08-25; 1779-08-31. 17 taken: *Lloyd's Evening Post*, 1779-08-18. 11 taken: *London Chronicle*, 1779-07-27. 10 taken: *Morning Chronicle and London Advertiser*, 1779-08-24. 2 taken: ibid., 1779-08-18. See also *Public Advertiser*, 1779-08-13; 1779-08-16. 10 taken: ibid., 1779-08-24. 25 taken: ibid., 1779-07-20. 19 taken: *St. James's Chronicle or the British Evening Post*, 1779-07-20.

84. *Morning Chronicle and London Advertiser*, 1780-08-15. *Public Advertiser*, 1780-07-04.

85. MB2, p. 86, 1781-06-28.

86. *Public Advertiser*, 1782-07-03. For the damage caused to the river's banks, see MB2, p. 154 (1782-07-01).

87. *Public Advertiser*, 1783-07-02.

88. *Parker's General Advertiser and Morning Intelligencer*, 1783-07-08.

89. *London Chronicle*, 1783-07-08. *Public Advertiser*, 1783-07-08.

90. MB2, p. 189, 1783-07-17.

91. *Public Advertiser*, 1783-07-15.

92. MB2, p. 189 (1783-07-24).

93. *Public Advertiser*, 1783-07-19; 1783-08-05; 1783-08-06.

94. *Gazetteer and New Daily Advertiser*, 1783-08-01.

95. *The Times*, 1785-21-07. *Public Advertiser*, 1784-05-14; 1784-05-12; 1784-05-19; 1784-07-12; 1786-06-12. *Morning Post and Daily Advertiser*, 1786-06-30. *General Evening Post*, 1786-06-29.

96. *Public Advertiser*, 1786-07-10. For indictments, see MB2, p. 215 (1784-27-05).

97. *Lloyd's Evening Post*, 1793-07-08. *Public Advertiser*, 1793-08-06. *True Briton*, 1794-29-08. *The Times*, 1797-08-18. *London Chronicle*, 1798-06-05. *Johnson's British Gazette and Sunday Monitor*, 1802-10-03. *Bell's Weekly Messenger*, 1809-12-12. *The Ipswich Journal*, 1809-12-16. *The Morning Post*, 1809-02-09. See also NR letters, LMA ACC/2558/ MW/C/15/142–48 and LMA ACC/2558/MW/C/15/105/1, cited in R. Ward 2003, 97.

98. On subscriptions and committees to prevent bathing, see MB3, p. 17 (1786-07-20); MB4, p. 101 (1794-08-07); MB6, p. 280 (1808-08-04). On the use of constables and threats, see MB3, p. 46 (1787-04-19); MB3, p. 47 (1787-04-26); MB4, p. 236 (1797-10-19); MB5, p. 155 (1801-08-13); MB6, p. 76 (1805-10-10). The problem reemerged after 1810 and was really solved only when the company began to supply free water at public baths. Ibid.

99. Sheard 2000, 63-86.

Chapter 8 · Epilogue and Legacy, 1800–1820

1. LBWW Minute book 6, LMA ACC/2558/LB/01/006 (1807-11-13).

2. R. Ward 2003, 161.

3. Hayman 2005, 64. Berg 1994, 46–47. Clark and Jacks 2007, 57.

4. See chapter 4.

5. Report to New River board 1814, LMA ACC/2558/NR/14/002, p. 18 (1814-03).

6. Hillier 2011.

7. Von Tunzelmann 1978.

8. Trew 2010. Pearson and Richardson 2001, 659–60. Harris, 2000 #611.

9. R. Ward 2003, 167.

10. Weale 1820, 21.

11. Rudden 1985, 135–39.

12. Report to New River board 1814, LMA ACC/2558/NR/14/002, pp. 1–3 (1814-03).

13. R. Ward 2003, ch. 15. Matthews 1835, 328–35.

14. Tynan 2002, 353–55. R. Ward 2003, ch. 15. Dickinson 1954, 89. Rudden 1985, 141–49.

15. Auxiron 1765, 55.

16. Vachette and Vachette 1791 [1797], 6. *Prospectus de la fourniture et distribution des eaux de la Seine, à Paris, par les machines à feu* 1781, 1–6.

17. Vachette and Vachette 1791 [1797], 6.

18. Bouchary 1946, 44. *Prospectus de la fourniture et distribution des eaux de la Seine, à Paris, par les machines à feu* 1781, 1–6.

19. Beaumont-Maillet 1991, ch. 10. Graber 2009. Bruyère 1804, 8–9.

20. D.A.V. 1823. Chatzis 2010. Guillerme 1988. Vienna built its network from 1841, http://www.wien.gv.at/wienwasser/versorgung/geschichte/. Berlin had no central supply until the 1850s: Mohajeri 2005, ch. 4. In 1856, 442,000 inhabitants used around ten thousand fountains (ibid., p. 28). There were some wooden pipes in place as early as

1560, but these rotted away and were never replaced (p. 37). The first proposals were based on studying systems in London and Paris (p. 40). The first water company was the Berlin Waterworks Company (1853), founded by Sir Charles Fox and Sir Thomas Russel Crampton (p. 50).

21. Chatzis 2010, 210.

22. "imité de celui adopté en Angleterre et notamment à Londres." Archives de l'École des ponts et chaussées, Manuscript 2749, cited in Chatzis 2010, 211. Mallet 1830. Mille 1854.

23. Mallet 1830, 9.

24. Chatzis 2010, 211. Mille 1854, 6.

25. Reden 1851, vol. 1, part 2, 1528.

26. Moeck-Schlömer 1998, 320–22.

27. Moeck-Schlömer 1998, 322–26.

28. Moeck-Schlömer 1998, 327–36.

29. Żelichowski 2004.

30. Meng 1993, 54.

31. Meng 1993, 40ff. Lindley and Mylne 1844.

32. Meng 1993, 61–63.

33. Reden 1851, vol. 1, part 2, 1528.

34. Warner 1968, 103–4. Melosi 2000, 19–23. Meyer 2006, 39–41. Chrimes 2002b.

35. Tomory 2012.

36. Israel 1998. Hughes 1983.

37. Daunton 2000, 5.

38. Great Britain, Poor Law Commission. 1842, 69-71.

39. Hamlin 1992, 683. Lewis 1952, 54–55. Melosi 2000, 44-52.

40. F. Ward 1880. See also F. Ward 1850.

41. Goldman 1997, 5. Benidickson 2007, 113.

Conclusion

1. Jun 1994. Hooimeijer 2008.

2. M. Jacob 1997. M. Jacob and Stewart 2004. M. Jacob 2014.

3. Nuvolari 2004. R. Allen 1983.

4. Harris 2000 made this argument for the East India Company and later trade companies.

5. Rosenberg and Birdzell 1986. Micklethwait and Wooldridge 2003. Landes 1969. For a counterpoint, see Guinnane et al. 2007.

6. MacLeod 1988. MacLeod 2009. Bottomley 2014. North and Thomas 1973, 155–56.

Archival Sources

British Library

Manuscripts Add 5063, Add 29974A, Add 34601, Add 70341, Harley 3604, Lansdowne 846, Stowe 747

J. Stafford and others, appellants. Mayor, commonalty, &c. of London, respondents. The Appellants case. 1719. Shelfmark 19.h.1.(140.)

London Metropolitan Archives

Chelsea Waterworks (ACC/2558/CH/01–02)
Corporation of London
    Alchin Collection (COL/AC/06)
    Chamberlain's Department: Rents and Rentals (COL/CHD/RN/02/01/026)
    Comptroller and City Solicitor: Solicitor (COL/CCS/SO)
    Journal of the Common Council (COL/CC/01/01)
    New River and Thames (COL/SJ/16/026)
    Repertory of the Court of Aldermen (COL/CA/01/01)
London Bridge Waterworks (ACC/2558/LB/01–02)
Metropolitan Water Board
    Clerk's Department (ACC/2558/MW/C/15/002–351)
    Survey Department (ACC/2558/MW/SU/01/040/2050)
New River Company (ACC/2558/NR/01–14)
    Minutes of the court of directors (ACC/2558/NR/01/1–9)
New River Company (ACC/1953/A/258)
Shadwell Waterworks (ACC/2558/S/A/01)

National Archives

Chancery (C 5, 225, 257)
New River Company (LR 2/27–44)
Privy Council (PC 2, 4, 12, 29)
Treasury Board (T 1)
Records of the Exchequer (E 178/6032)
State Papers (SP 9, 14, 29, 59)

Nottingham University

Harley family papers (Pw2 Hy)

Primary and Secondary Sources

An Act concerning the repairing, making, and amending of the conduits in London. 1543.

Acts of the Privy Council of England. New Series. 1927. London: H.M.S.O.

Adolphus, John. 1818. *The political state of the British Empire containing a general view of the domestic and foreign possessions of the crown, the laws, commerce, revenues, offices, and other establishments, civil and military.* 4 vols. London: printed for T. Cadell and W. Davies.

Aguilar, Ephraim Lópes Pereira, Baron d'. 1906. In *The Jewish encyclopedia: a descriptive record of the history, religion, literature, and customs of the Jewish people from the earliest times to the present day*, ed. C. Adler and I. Singer. New York: Funk & Wagnalls.

Albert, William, Michael J. Freeman, and Derek Howard Aldcroft, eds. 1983. *Transport in the industrial revolution.* Manchester: Manchester University Press.

Allen, Michelle Elizabeth. 2008. *Cleansing the city: sanitary geographies in Victorian London.* Athens: Ohio University Press.

Allen, Robert C. 1983. "Collective invention." *Journal of Economic Behavior and Organization* 4: 1–24.

——. 2009. *The British industrial revolution in global perspective.* Cambridge: Cambridge University Press.

Alsford, Stephen. 2011. *Florilegium Urbanum: The great conduit.* http://users.trytel.com /~tristan/towns/florilegium/community/cmfabr24.html.

Anderson, Adam. 1764. *An historical and chronological deduction of the origin of commerce: from the earliest accounts, containing an history of the great commercial interests of the British Empire.* London: A. Millar.

Archenholz, Johann Wilhelm von. 1785. *England und Italien.* Leipzig: Im Verlage der Dykischen Buchhandlung.

Aubrey, John, and Andrew Clark. 1898. *Brief lives, chiefly of contemporaries, set down by John Aubrey, between the years 1669 & 1696.* 2 vols. Oxford: Clarendon Press.

Auxiron, Claude-François-Joseph. 1765. *Projet patriotique sur les eaux de Paris, ou, Mémoire sur les moyens de fournir à la ville de Paris des eaux saines.*

B. R. 1730. *A new view, and observations on the ancient and present state of London and Westminster. With an account of the most remarkable accidents as to wars, fires, plagues, and other occurences.* London: printed for A. Bettesworth and C. Hitch.

Baker, Henry. 1743. *An attempt towards a natural history of the polype in a letter to Martin Folkes, Esq.* London: printed for R. Dodsley, at Tully's Head in Pall-Mall, and sold by M. Cooper in Pater-Noster-Row, and J. Cuff, optician, in Fleetstreet.

Baldwin, Robert. 2002. "Bulmer, Sir Bevis." In *A biographical dictionary of civil engineers in Great Britain and Ireland*, vol. 1: *1500–1830*, ed. A. W. Skempton. London: T. Telford.

Barber, Peter. 1992. "England II: monarchs, ministers, and maps, 1550–1625." In *Monarchs, ministers, and maps: the emergence of cartography as a tool of government in early modern Europe*, ed. D. Buisseret, 57–98. Chicago: University of Chicago Press.

Bate, John. 1635. *The mysteries of nature and art: in foure severall parts.* 2nd ed. London: printed for Ralph Mabb.

Beaumont-Maillet, Laure. 1991. *L'eau à Paris.* Paris: Hazan.

Beier, A. L. 1986. "Engine of manufacture: the trades in London." In *London 1500–1700: the making of the metropolis*, ed. A. L. Beier and R. Finlay, 141–67. London: Longman.

Beighton, Henry. 1731. "A description of the water-works at London-Bridge, explaining the draught of tab. I." *Philosophical Transactions of the Royal Society of London* 37: 5–12.

Benidickson, Jamie. 2007. *The culture of flushing: a social and legal history of sewage.* Vancouver: UBC Press.

Berg, Maxine. 1994. *The age of manufactures, 1700–1820: industry, innovation, and work in Britain.* 2nd ed. New York: Routledge.

Bernardoni, Andrea. 2008. "Tecniche e macchine per l'alesatura delle artiglierie nel Rinascimento." *Le Journal de la Renaissance* 6: 201–17.

———. 2009. "The biography of a technical picture: Biringuccio's boring machine woodcut." *Nuncius* 24: 291–311.

Berry, G. C. 1956. "Sir Hugh Myddelton and the New River." *Transactions of the Honourable Society of Cymmrodorion*: 17–46.

Bevan, G. Phillips. 1884. *The London water supply; its past, present, and future.* London: Stanford.

Blackstone, Geoffrey Vaughan. 1957. *A history of the British fire service.* London: Routledge and K. Paul.

Bocquet, Denis, Konstantinos Chatzis, and Agnès Sander. 2008. "From free good to commodity: universalizing the provision of water in Paris (1830–1940)." *Geoforum* 39: 1821–32.

Bogart, Dan. 2011. "Did the Glorious Revolution contribute to the transport revolution? Evidence from investment in roads and rivers." *Economic History Review* 64: 1073–1112.

Bond, C. J. 1993. "Water management in the urban monastery." In *Advances in monastic archaeology*, ed. R. Gilchrist and H. Mytum, 43–78. Oxford: British Archaeological Research.

Bonneville, Françoise de. 1998. *The book of the bath.* New York: Rizzoli.

Borsay, Peter. 2001. "London, 1660–1800: a distinctive culture?" In *Two capitals: London and Dublin, 1500–1840*, ed. P. Clark and R. Gillespie, 167–84. Oxford: published for the British Academy by Oxford University Press.

Bossut, Charles. 1786. *Traité theórique et expérimental d'hydrodynamique.* 2 vols. Paris: De l'Imprimerie royale.

Bottomley, Sean. 2014. *The British patent system and the industrial revolution, 1700–1852: from privilege to property.* Cambridge: Cambridge University Press.

Bouchary, Jean. 1946. *La Compagnie des Eaux de Paris et l'entreprise de l'Yvette: l'eau à Paris à la fin du XVIIIe siècle.* Paris: Rivière.

Boulton, Jeremy. 2000. "London 1540–1700." In *The Cambridge urban history of Britain*, vol. 2: *1540–1840*, ed. P. Clark, 315–46. Cambridge: Cambridge University Press.

Broadberry, Stephen, Bruce Campbell, Alexander Klein, Mark Overton, and Bas van Leeuwen. 2014. *British economic growth, 1270–1870.* Cambridge: Cambridge University Press.

Brown, Josiah. 1779. *Reports of cases, upon appeals and writs of error in the High Court of Parliament; from the year 1701, to the year 1779.* Vol. 2. London: printed by His Majesty's Law-Printers.

Brown, Michael. 2008. "From foetid air to filth: the cultural transformation of British epidemiological thought, ca. 1780–1848." *Bulletin of the History of Medicine* 82: 515–44.

Bruyère, Louis. 1804. *Rapport du 9 floréal an X, sur les moyens de fournir l'eau nécessaire à la ville de Paris, et particulièrement sur la dérivation des rivières d'Ourcq, de la Beuvronne, de l'Yvette, de la Bièvre et autres, par L. Bruyère.* Paris: Courcier.

The builder's dictionary or, Gentleman and Architect's companion. 1734. London: printed for A. Bettesworth and C. Hitch.

Butler, Stella. 2004. "White, Charles." In *Oxford dictionary of national biography*, ed. H. C. G. Matthew, B. Harrison, and L. Goldman. Oxford: Oxford University Press.

Calendar of State Papers Domestic. 1858–1972. London: H.M.S.O. british-history.ac.uk.

Carbonnier, Youri. 2006. *Maisons parisiennes des Lumières.* Paris: Presses de l'Université Paris-Sorbonne.

Carr, Cecil T. 1913. *Select charters of trading companies, A.D. 1530–1707.* The publications of the Selden Society, vol. 28. London: B. Quaritch.

Carruthers, Bruce G. 1996. *City of capital politics and markets in the English financial revolution.* Princeton, N.J.: Princeton University Press.

Castle, Richard. 1735. *An essay toward supplying the city of Dublin with water.* Dublin: printed by Syl. Pepyat, Printer to the Honourable City of Dublin.

——. 2012. In *Dictionary of Irish architects, 1720–1940.* Dublin: Irish Architectural Archive.

A catalogue of the Harleian collection of manuscripts, purchased by authority of Parliament, for the use of the publick; and preserved in the British Museum. 1759. London: printed by Dryden Leach, and sold by L. Davis and C. Reymers, opposite Grays-Inn, Holborn.

Chandler, Alfred D. 1977. *The visible hand: the managerial revolution in American business.* Cambridge, Mass.: Belknap Press.

Chapman, Colin R. 1994. *Pre-1841 censuses & population listings in the British Isles.* 4th ed. Chapmans records cameos series. Baltimore: Genealogical Publishing.

Chartres, John. 1986. "Food consumption and internal trade." In *London, 1500–1700: the making of the metropolis*, ed. A. L. Beier and R. Finlay, 168–96. London: Longman.

Chatzis, Konstantinos. 2010. "Eaux de Paris, eaux de Londres: quand les ingénieurs de la capitale française regardent outre-Manche, 1820–1880." *Documents pour l'histoire des techniques* 19: 209–18.

Chatzis, Konstantinos, and Olivier Coutard. 2005. "Water and gas: early developments in the utility networks of Paris." *Journal of Urban Technology* 12: 1–17.

Chomel, Noel. 1725. *Dictionaire oeconomique: or, The family dictionary.* London: printed for D. Midwinter.

Chrimes, Mike. 2002a. "Foulds, John Torr." In *A biographical dictionary of civil engineers in Great Britain and Ireland*, vol. 1: *1500–1830*, ed. A. W. Skempton. London: T. Telford.

——. 2002b. "Latrobe, Benjamin Henry." In *A biographical dictionary of civil engineers in Great Britain and Ireland*, vol. 1: *1500–1830*, ed. A. W. Skempton. London: T. Telford.

Clark, Gregory. 1996. "The political foundations of modern economic growth: England, 1540–1800." *Journal of Interdisciplinary History* 26: 563–88.

——. 2004. "The price history of English agriculture, 1209–1914." *Research in Economic History* 22: 41–123.

Clark, Gregory, and David Jacks. 2007. "Coal and the industrial revolution, 1700–1869." *European Review of Economic History* 11: 39–72.

Clifford, Frederick. 1885. *A history of private bill legislation.* London: Butterworth's.

Cockayne, Emily. 2007. *Hubbub: filth, noise, & stench in England, 1600–1770*. New Haven, Conn.: Yale University Press.

Coffman, D'Maris. 2013. *Questioning credible commitment: perspectives on the rise of financial capitalism*. Cambridge: Cambridge University Press.

Coley, Noel G. 1979. "'Cures without care': 'Chymical physicians' and mineral waters in seventeenth-century English medicine." *Medical History* 23: 191–214.

———. 1982. "Physicians and the chemical analysis of mineral waters in eighteenth-century England." *Medical History* 26: 123–44.

———. 1990. "Physicians, chemists and the analysis of mineral waters: 'the most difficult part of chemistry.'" *Medical History Supplement (The Medical History of Waters and Spas)* 10: 56–66.

Colston, James. 1890. *The Edinburgh and District water supply. A historical sketch*. Edinburgh: Colston.

Colvin, Howard, ed. 1982. *The history of the King's works*. Vol. 4: *1485–1660 (Part II)*. London: H.M.S.O.

Concanen, Matthew, and Aaron Morgan. 1795. *The history and antiquities of the parish of St. Saviour's, Southwark*. London: printed by J. Delahoy.

Cook, G. C. 2001. "Construction of London's Victorian sewers: the vital role of Joseph Bazalgette." *Postgraduate Medical Journal Online* 77: 802–4.

Coutard, Olivier, ed. 1999. *The governance of large technical systems*. London: Routledge.

Coyer, Gabriel-François. 1783. *Oeuvres de M. L'Abbé Coyer, des académies de Nancy, de Rome, et de Londres*. Paris: Duchesene.

Cramsie, John. 2000. "Commercial projects and the fiscal policy of James VI and I." *Historical Journal* 43: 345–64.

Culpeper, Nicholas. 1652. *Catastrophe magnatum, or, The fall of monarchie a caveat to magistrates, deduced from the eclipse of the sunne, March 29, 1652, with a probable conjecture of the determination of the effects*. London: printed for T. Vere and Nath. Brooke.

The curious modern Traveller being some late and particular observations on the greatest Curiosities and Antiquities in Most Parts of the World: Especially in Great Britain, Ireland, France & c. London. 1745. London: printed by A. M'Culloh.

D'Acres, R. 1660. *The art of water-drawing, or A compendious abstract of all sorts of water-machins, or gins, practised in the world with their natural grounds and reasons*. London: printed for Henry Brome, at the Gun in Ivie-Lane.

Daunton, Martin J. 2000. Introduction to *The Cambridge urban history of Britain*, vol. 3: *1840–1950*, ed. P. Clark, D. M. Palliser, and M. J. Daunton, 1–58. Cambridge: Cambridge University Press.

D.A.V. 1823. *Projet de distribution générale dans l'intérieur de Paris, de 7.678 modules, ou 4.000 ponces des eaux de l'Ourcq, suivant le Système adopté en Angleterre, et notamment à Londres*. Paris: imp. de Fain.

Defoe, Daniel. 1725. *A tour thro' the whole island of Great Britain: Divided into circuits or journeys, giving a particular and entertaining account of whatever is curious and worth pobservation*. London: printed and sold by G. Strahan, in Cornhill.

———. 1887. *An essay upon projects*. London: Cassell.

De Laune, Thomas, and John Stow. 1681. *The present state of London, or, Memorials comprehending a full and succinct account of the ancient and modern state thereof*. London: printed by George Larkin.

Desaguliers, J. T. 1726. "An account of several experiments concerning the running of water in pipes, as it is retarded by friction and intermixed air, some of which were made before the Royal Society on Thursday the 5th of May, 1726." *Philosophical Transactions (1683–1775)* 34: 77–82.

———. 1744. *A course of experimental philosophy.* 2 vols. London: W. Innys.

A description of the county of Middlesex containing a circumstantial account of its public buildings, seats of the nobility and gentry, places of resort and entertainment, Curiosities of Nature and Art, (including those of London and Westminster). 1775. London: printed for R. Snagg.

De Vries, Jan. 1984. *European urbanization, 1500–1800.* Cambridge, Mass.: Harvard University Press.

———. 2008. *The industrious revolution: consumer behavior and the household economy, 1650 to the present.* Cambridge: Cambridge University Press.

Dézallier d'Argenville, A. J. 1747. *La theorie et la pratique du jardinage. Ou l'on traite a fond des beaux jardins apellés communément les jardins de plaisance et de propreté.* Paris: Pierre-Jean Mariette.

Dickinson, H. W. 1954. *Water supply of Greater London, being a series of articles originally published in The engineer in 1948.* Leamington Spa: printed for the Newcomen Society at the Courier Press.

Dickson, P. G. M. 1967. *The financial revolution in England: a study in the development of public credit, 1688–1756.* London: Macmillan; New York: St. Martin's.

Dodd, Arthur Herbert. 2014. "Myddelton family." In *Welsh Biography Online.* yba.llgc.org .uk/en/s3-MYDD-ELT-1207.html.

Donaldson, Thomas. 1982. *Corporations and morality.* Englewood Cliffs, N.J.: Prentice-Hall.

Douglas, Mary. 1966. *Purity and danger: an analysis of concepts of pollution and taboo.* New York: Praeger.

Downes, Kerry. 2004. "Wren, Sir Christopher (1632–1723)." In *Oxford dictionary of national biography,* ed. H. C. G. Matthew, B. Harrison, and L. Goldman. Oxford: Oxford University Press.

Du Buat, Chevalier. 1786. *Principes d'hydraulique, vérifiés par un grand nombre d'expériences faites par ordre du gouvernement: ouvrage dans lequel on traite du mouvement uniforme & varié de l'eau dans les rivières.* 2nd ed. Paris: Impr. de Monsieur.

Duffy, Christopher. 1979. *Siege warfare: the fortress in the early modern world, 1494–1660.* Vol. 1. London: Routledge.

Earle, Peter. 1989. *The making of the English middle class: business, society, and family life in London, 1660–1730.* Berkeley: University of California Press.

———. 2001. "The economy of London, 1660–1730." In *Urban achievement in early modern Europe: golden ages in Antwerp, Amsterdam, and London,* ed. P. O'Brien, 81–96. Cambridge: Cambridge University Press.

Elias, Norbert. 1994. *The civilizing process: the history of manners and state formation and civilization.* Oxford: Blackwell.

Ellis, William. 1734. *The London and country brewer.* London: printed and sold by W. Meadows.

———. 1736. *The london and country brewer.* 2nd ed., corrected. London: printed for Messieurs Fox, at the Half-Moon and Seven Stars, in Westminster-Hall.

Entick, John. 1766. *A new and accurate history and survey of London Westminster, Southwark, and places adjacent*. London: printed for Edward and Charles Dilly.

Evans, Robert. 1987. "The early history of fire insurance." *Journal of Legal History* 8: 88–91.

Extracts from the books of the Mayor and Aldermen of Hertford together with copies of papers in their custody, relating to the navigation on the River Lea, between Hertford and Ware. 1734. London.

Fabricius, Johann Christian. 1784. *Briefe aus London vermischten Inhalts*. Dessau: Buchhandlung der Gelehrten.

Fairclough, K. R. 2004. "Sorocold, George (c.1668–1738?)." In *Oxford dictionary of national biography*, ed. H. C. G. Matthew, B. Harrison, and L. Goldman. Oxford: Oxford University Press.

Farey, John. 1827. *A treatise on the steam engine: historical, practical, and descriptive*. London: printed for Longman, Rees, Orme, Brown and Green.

Faulkner, Thomas. 1820. *History and antiquities of Kensington*. London: D. Jaques.

Ford, Edward, and W. Roberts. 1641. *A design for bringing a navigable river from Rickmansworth in Hartfordshire to St. Giles's in the Fields the benefits of it declared, and the objections answered*. London: printed for John Clarke.

Fothergill, John, and John Coakley Lettsom. 1784. *The works of John Fothergill*. London: Charles Dilly.

Freeman, Mark, Robin Pearson, and James Taylor. 2012. *Shareholder democracies? corporate governance in Britain and Ireland before 1850*. Chicago: University of Chicago Press.

Gatrell, V. A. C. 2007. *City of laughter: sex and satire in eighteenth-century London*. New York: Walker.

Geels, Frank. 2005. "Co-evolution of technology and society: the transition in water supply and personal hygiene in the Netherlands (1850–1930)—a case study in multi-level perspective." *Technology in Society* 27: 363–97.

Getzler, Joshua. 2004. *A history of water rights at common law*. Oxford: Oxford University Press.

Girard, Pierre-Simon. 1812. *Recherches sur les eaux publiques de Paris, les distributions successives qui ont été proposés pour en augmenter le volume*. Paris: Imprimerie Impériale.

Goddard, Nicholas, and John Sheail. 2004. "Victorian sanitary reform: where were the nnovators?" In *Environmental problems in European cities in the 19th and 20th century*, ed. C. Bernhardt, 87–103. Münster: Waxmann Verlag.

Goldman, Joanne Abel. 1997. *Building New York's sewers: developing mechanisms of urban management*. West Lafayette, Ind.: Purdue University Press.

Goldsmith, Hugh, and Dan Carter. 2015. "Financing the evolution of London's water services: 1582 to 1904." *Milan European Economy Workshops*. Working Paper no. 2015-02.

Gosling, William. 1643. *Seasonable advice, for preventing the mischiefe of fire, that may come by negligence, treason, or otherwise Ordered to be printed by the Lord Major of London*. London: printed for H.B. at the Castle in Corn-hill.

Gough, J. W. 1964. *Sir Hugh Myddleton: entrepreneur and engineer*. Oxford: Clarendon Press.

Gould, Robert. 1689. *Poems, chiefly consisting of satyrs and satyrical epistles*. London.

Graber, Frédéric. 2009. *Paris a besoin d'eau: projet, dispute et délibération technique dans la France napoléonienne*. Paris: CNRS.

Great Britain. Poor Law Commission. 1842. *Report to Her Majesty's Principal Secretary of State for the Home Department, from the Poor Law Commissioners, on an inquiry into the sanitary condition of the labouring population of Great Britain, with appendices*. London: printed by Clowes for H.M.S.O.

Green, Ephraim. 1717. *The case of Ephraim Green; late clerk to the New-River Company, and surveyor of their works*. London.

Grewe, Klaus. 1991. *Wasserversorgung und -entsorgung im Mittelalter. Ein technik-geschichtlicher Überblick*. Mainz: Philipp von Zabern.

——. 2000. "Water technology in medieval Germany." In *Working with water in medieval Europe: technology and resource-use*, ed. P. Squatriti, 129–60. Leiden: Brill.

Grimm, Johann Friedrich Karl. 1775. *Bemerkungen eines reisenden durch Deutschland, Frankreich, England und Holland in briefen an seine freunde*. 3 vols. Altenburg: In der Richterischen Buchhandlung.

Guillerme, André. 1988. "The genesis of water supply, distribution, and sewerage systems in France, 1800–1850." In *Technology and the rise of the networked city in Europe and America*, ed. J. A. Tarr and D. Gabriel, 91–115. Philadelphia: Temple University Press.

Guinnane, Timothy, Ron Harris, Naomi R. Lamoreaux, and Jean-Laurent Rosenthal. 2007. "Putting the corporation in its place." *Enterprise and Society* 8: 687–729.

Haigh, James. 1779. *A hint to the dyers, and cloth-makers and well worth the notice of the merchant*. London: printed for Mess. Rivington.

Hall, A. Rupert. 1974. "What did the industrial revolution in Britain owe to science?" In *Historical perspectives: studies in English thought and society, in honour of J. H. Plumb*, ed. N. McKendrick, 129–51. London: Europa.

Halliday, Stephen. 1999. *The great stink of London: Sir Joseph Bazalgette and the cleansing of the Victorian capital*. Stroud: Sutton.

Hamlin, Christopher. 1990. *A science of impurity: water analysis in nineteenth century Britain*. Berkeley: University of California Press.

——. 1992. "Edwin Chadwick and the Engineers, 1842–1854: systems and antisystems in the pipe-and-brick sewers war." *Technology and Culture* 33: 680–709.

——. 1994. "State medicine in Great Britain." In *The history of public health and the modern state*, ed. D. Porter, 132–64. Amsterdam: Editions Rodopi.

——. 1998. *Public health and social justice in the age of Chadwick: Britain, 1800–1854*. Cambridge. Cambridge University Press.

——. 2000. "Water." In *The Cambridge world history of food*, ed. K. F. Kiple and K. C. Ornelas, 720–30. Cambridge: Cambridge University Press.

——. 2012. Introduction to *Sanitary reform in Victorian Britain. Part 1*, ed. T. Y. Choi, C. Hamlin, and M. Allen-Emerson, vii–xiii. London: Pickering & Chatto.

Hancocke, John. 1723. *Febrifugum magnum or, common water the best cure for fevers, and probably for the plague*. 4th ed. London: printed for R. Halsey.

Hardy, Anne. 1984. "Water and the search for public health in London in the eighteenth and nineteenth centuries." *Medical History* 28: 250–82.

Harris, Ron. 1994. "The Bubble Act: its passage and its effects on business organization." *Journal of Economic History* 54: 610–27.

———. 2000. *Industrializing English law: entrepreneurship and business organization, 1720–1844*. Cambridge: Cambridge University Press.

Harvey, James. 1701. *Scelera aquarum: or, a supplement to Mr. Graunt on the bills of mortality. Shewing as well the causes, as encrease of the London, Parisian, and Amsterdam scorbute*. London: printed for the author, and sold by Du Chemin.

Hassan, John. 1985. "The growth and impact of the British water industry in the nineteenth century." *Economic History Review* 38: 531–47.

———. 1998. *A history of water in modern England and Wales*. Manchester: Manchester University Press.

Hatton, Edward. 1708. *A new view of London; or, an ample account of that city*. London: printed for R. Chiswell.

Hayman, Richard. 2005. *Ironmaking: the history and archaeology of the iron industry*. Stroud: Tempus.

Heberden, Ernest. 1986. "William Heberden the elder (1710–1801): aspects of his London practice." *Medical History* 30: 303–21.

———. 2004. "Heberden, William." In *Oxford dictionary of national biography*, ed. H. C. G. Matthew, B. Harrison, and L. Goldman. Oxford: Oxford University Press.

Heberden, William. 1768. "Remarks on the pump-water of London, and on the methods of procuring the purest water." *The annual register, or, A view of the history, politicks, and literature for the year 1768* 11: 86–91.

Hembry, Phyllis. 1990. *The English spa, 1560–1815: a social history*. London: Athlone Press.

Herring, Francis. 1625. *Certaine rules, directions, or advertisements for this time of pestilentiall contagion*. London: printed by William Iones.

Hill, Donald Routledge. 1984. *A history of engineering in classical and medieval times*. London: Croom Helm.

Hillier, Joseph. 2011. "The rise of constant water in nineteenth-century London." *London Journal* 36: 37–53.

Holinshed, Raphael. 1587. *The Third volume of Chronicles, beginning at duke William the Norman, commonlie called the Conqueror; and descending by degrees of yeeres to all the kings and queenes of England in their orderlie successions*. London: printed by Henry Denham.

Hooimeijer, Fransje. 2008. "History of urban water in Japan." In *Urban water in Japan*, ed. R. de Graaf and F. Hooimeijer, 17–86. London: Taylor & Francis.

Hooke, Robert, and W. Derham. 1726. *Philosophical experiments and observations of the late eminent Dr. Robert Hooke, S.R.S. and Geom. Prof. Gresh. and other eminent virtuoso's in his time*. London: printed by W. and J. Innys.

Hoppit, Julian. 2000. *A land of liberty? England, 1689–1727*. Oxford: Oxford University Press.

———. 2011. "Compulsion, compensation and property rights in Britain, 1688–1833." *Past & Present* 210: 93–128.

———. 2014. "Political power and British economic life, 1650–1870." In *The Cambridge economic history of modern Britain*, vol. 1: *1700–1860*, ed. R. Floud, J. Humphries, and P. Johnson, 344–67. Cambridge: Cambridge University Press.

Horrell, Sara. 2014. "Consumption, 1700–1870." In *The Cambridge economic history of modern Britain*, vol. 1: *1700–1860*, ed. R. Floud, J. Humphries, and P. Johnson, 237–64. Cambridge: Cambridge University Press.

Horseman, Gilbert. 1744. *Precedents in conveyancing, settled and approved.* London: printed by H. Lintot.

Howell, James. 1657. *Londinopolis: an historical discourse, Perlustration of the city of London, the imperial chamber, and chief emporium of Great Britain.* London: printed by J. Streater.

Hughes, Thomas Parke. 1983. *Networks of power: electrification in Western society, 1880–1930.* Baltimore: Johns Hopkins University Press.

Hughson, David. 1809. *London; being an accurate history and description of the British metropolis and its neighbourhood: to thirty miles extent, from an actual perambulation.* London: J. Stratford.

Hunter, Matthew C. 2013. *Wicked intelligence: visual art and the science of experiment in Restoration London.* Chicago: University of Chicago Press.

Hutton, Charles. 1795. *A mathematical and philosophical dictionary: containing an explanation of the terms, and an account of the several subjects, comprized under the heads mathematics, astronomy, and philosophy both natural and experimental.* London: printed by J. Davis.

Innes, Joanna. 2001. "Managing the metropolis: London's social problems and their control, c. 1660–1830." In *Two capitals: London and Dublin, 1500–1840,* ed. P. Clark and R. Gillespie, 53–80. Oxford: published for the British Academy by Oxford University Press.

Israel, Paul. 1998. *Edison: a life of invention.* New York: John Wiley.

Jacob, Giles. 1717. *The country gentleman's vade mecum. Containing an account of the best methods to improve lands, plowing and sowing of corn.* London: printed for William Taylor at the Ship in Pater-Noster-Row.

Jacob, Margaret C. 1997. *Scientific culture and the making of the industrial West.* Oxford: Oxford University Press.

———. 2014. *The first knowledge economy: human capital and the European economy, 1750–1850.* Cambridge: Cambridge University Press.

Jacob, Margaret C., and Larry Stewart. 2004. *Practical matter: Newton's science in the service of industry and empire, 1687–1851.* Cambridge, Mass.: Harvard University Press.

Jenkins, R. 1911. "Bevis Bulmer." *Notes and Queries* 4: 401–3.

———. 1928. "A chapter in the history of the water supply of London: A Thames-side pumping installation and Sir Edward Ford's patent from Cromwell." *Transactions of the Newcomen Society* 9: 43–51.

———. 1895. "Notes on the London Bridge Waterworks." *The Antiquary* 12: 243–46, 261–65.

Jenner, Mark S. R. 1996. "Quackery and enthusiasm, or why drinking water cured the plague." In *Religio medici: medicine and religion in seventeenth-century England,* ed. O. P. Grell and A. Cunningham, 313–40. Aldershot: Scolar Press.

———. 1998. "Bathing and baptism: Sir John Floyer and the politics of cold bathing." In *Refiguring revolutions: aesthetics and politics from the English revolution to the Romantic revolution,* ed. K. Sharpe and S. N. Zwicker, 197–216. Berkeley: University of California Press.

———. 2000. "From conduit community to commercial network? Water in London 1500–1725." In *Londinopolis: essays in the social and cultural history of early modern London,* ed. P. Griffiths and M. Jenner, 250–72. Manchester: Manchester University Press.

———. 2007. "Monopoly, markets and public health: pollution and commerce in the history of London water, 1780–1830." In *Medicine and the market in pre-modern England and its colonies, 1450–1850*, ed. M. Jenner and P. Wallis, 216–37. Basingstoke: Palgrave.

Journal of the House of Commons. 1767–1830. London: H.M.S.O. www.british-history.ac.uk.

Journal of the House of Lords. 1802. London: H.M.S.O. www.british-history.ac.uk.

Jun, Hatano. 1994. "Edo's water supply." In *Edo and Paris: urban life and the state in the early modern era*, ed. J. L. McCiain, J. M. Merrirnan, and U. Kaoru, 234–50. Ithaca, N.Y.: Cornell University Press.

Keene, Derek. 2001a. "Growth, modernisation, and control: the transformation of London's landscape c. 1500–c. 1760." In *Two capitals: London and Dublin, 1500–1840*, ed. P. Clark and R. Gillespie, 7–38. Oxford: published for the British Academy by Oxford University Press.

———. 2001b. "Issues of water in medieval London to c. 1300." *Urban History* 28: 161–79.

Kellett, J. R. 1963. "The financial crisis of the Corporation of London and the Orphans' Act, 1694." *Guildhall Miscellany* 2: 220–27.

King, Peter. 2005. "The production and consumption of bar iron in early modern England and Wales." *Economic History Review* 58: 1–33.

Kirby, Maurice W. 1994. "Big business before 1900." In *Business enterprise in modern Britain: from the eighteenth to the twentieth century*, ed. M. W. Kirby and M. B. Rose, 113–37. London: Routledge.

Knissel, Walter, and Gerhard Fleisch. 2004. *Kulturdenkmal "Oberharzer Wasserregal"-eine epochale Leistung*. Clausthal-Zellerfeld: Technische Universität Clausthal.

Krünitz, Johann Georg. 1801. "London." In *Oekonomische-Technologische Encyclopädie; oder, Allgemeines System der Land-, Haus-, und Staats-Wirthschaft*, ed. J. G. Krünitz. Berlin: J. Pauli.

Kucher, Michael P. 2004. *The water supply system of Siena, Italy: the medieval roots of the modern networked city*. New York: Routledge.

Landers, John. 1993. *Death and the metropolis: studies in the demographic history of London, 1670–1830*. Cambridge: Cambridge University Press.

Landes, David S. 1969. *The unbound Prometheus: technological change and industrial development in Western Europe from 1750 to the present*. Cambridge: Cambridge University Press.

Lewis, Richard Albert. 1952. *Edwin Chadwick and the public health movement, 1832–1854*. London: Longmans.

Ley, A. J. 2000. *A history of building control in England and Wales, 1840–1990*. Coventry: RICS Books.

Liessmann, Wilfried. 1992. *Historischer Bergbau im Harz: ein Kurzführer*. 2nd ed. Cologne: Sven von Loga.

Lindley, William, and William Chadwell Mylne. 1844. *Ingenieur-Bericht, die Anlage der Oeffentlichen Wasser-Kunst für die Stadt Hamburg betreffend*. London.

"London." 1829. In *London encyclopaedia; or, Universal dictionary of science, art, literature, and practical mechanics*, 137–255. London: T. Tegg.

London and its environs described containing an account of whatever is most remarkable for grandeur, elegance, curiosity or use, in the city and in the country twenty miles round it. In six volumes. 1761. London: printed for R. and J. Dodsley in Pall-Mall.

London County Council, George Henry Gater, E. P. Wheeler, and London Survey Committee. 1937. *The Strand (The parish of St. Martin-in-the-fields, part II), Survey of London*. London: published for the London County Council.

London lives, 1690 to 1800: Crime, poverty and social policy in the metropolis. 2015. www .londonlives.org.

Long, Pamela O. 2008. "Hydraulic engineering and the study of antiquity: Rome, 1557–70." *Renaissance Quarterly* 61: 1098–1138.

Lorrain, Dominique. 2005. "Gig@city: The rise of technological networks in daily life." In *Sustaining urban networks: the social diffusion of large technical systems*, ed. O. Coutard, R. Zimmerman, and R. E. Hanley, 15–31. London: Routledge.

Lowthorp, John. 1699. "An experiment of the refraction of the air made at the command of the Royal Society, Mar. 28. 1699." *Philosophical Transactions (1683–1775)* 21: 339–42.

Lysons, Daniel. 1795. *The environs of London being an historical account of the towns, villages, and hamlets, within twelve miles of that capital*. London: printed by A. Strahan.

Macleod, Christine. 1986. "The 1690s patents boom: invention or stock-jobbing?" *Economic History Review* 39: 549–71.

———. 1988. *Inventing the industrial revolution: the English patent system, 1660–1800*. Cambridge: Cambridge University Press.

———. 2009. "Patents for invention? Setting the stage for the British industrial revolution?" *Empiria: Revista de metodología de ciencias sociales* 18: 37–58.

Magnusson, Roberta J. 2001. *Water technology in the Middle Ages: cities, monasteries, and waterworks after the Roman Empire*. Baltimore: Johns Hopkins University Press.

Maitland, William. 1739. *The history of London from its foundation by the Romans, to the present time*. London: printed by Samuel Richardson.

———. 1756. *The history of London from its foundation to the present time*. London: printed for T. Osborne and J. Shipton.

Malanima, Paolo. 2010. "Urbanization." In *The Cambridge economic history of modern Europe, 1700–1870*, vol. 1, ed. S. Broadberry and K. O'Rourke, 235–63. Cambridge: Cambridge University Press.

Mallet, Charles-François. 1830. *Notice historique sur le projet d'une distribution générale d'eau à domicile dans Paris et exposé de détails y relatifs, recueillis dans différentes villes du Royaume-Uni, notamment à Londres*. Paris: Carilian-Goeury.

Marshall, John. 1832. *Mortality of the metropolis a statistical view of the number of persons reported to have died, of each of more than 100 kinds of disease and casualties within the bills of mortality, in each of the two hundred and four years, 1629–1831*. London: printed for J. Marshall.

Martin, Benjamin. 1759. *The natural history of England; or, a description of each particular county*. London: printed by W. Owen.

Martin, Thomas. 1813. *The circle of the mechanical arts containing practical treatises on the various manual arts, trades, and manufactures*. 2nd ed. London: printed for Richard Rees.

Massounie, Dominique. 2009. *Les monuments de l'eau: aqueducs, châteaux d'eau et fontaines dans la France urbaine, du règne de Louis XIV à la révolution*. Temps & espace des arts. Paris: Patrimoine.

Masters, Betty R., ed. 1984. *Chamber accounts of the sixteenth century*. London: London Record Society for the Corporation of London.

Mathias, Peter. 1959. *The brewing industry in England, 1700–1830*. Cambridge: Cambridge University Press.

Matthews, William. 1835. *Hydraulia, an historical and descriptive account of the water works of London: and the contrivances for supplying other great cities, in different ages and countries*. London: Simpkin Marshall.

Maxwell, Lee M. 2003. *Save womens lives: history of washing machines*. Eaton, Colo.: Oldewash.

Mayntz, Renate, and Thomas Parke Hughes. 1988. *The development of large technical systems*. Frankfurt am Main: Campus Verlag.

McKellar, Elizabeth. 1999. *The birth of modern London: the development and design of the city, 1660–1720*. Manchester: Manchester University Press.

McKendrick, Neil. 1974. "Home demand and economic growth: a new view of the role of women and children in the industrial revolution." In *Historical perspectives: studies in English thought and society in honor of J. H. Plumb*, ed. N. McKendrick, 152–210. London: Europa.

McKendrick, Neil, John Brewer, and J. H. Plumb. 1982. *The birth of a consumer society: the commercialization of eighteenth-century England*. Bloomington: Indiana University Press.

Melosi, Martin V. 2000. *The sanitary city: urban infrastructure in America from colonial times to the present*. Baltimore: Johns Hopkins University Press.

"Memoirs of Mr. William Lamb, Founder of Lamb's Chapel, etc." 1783. *Gentleman's Magazine and Historical Chronicle* 53.

Meng, Alfred. 1993. *Geschichte der Hamburger Wasserversorgung*. Hamburg: Medien-Verl. Schubert.

Merritt, J. F. 2001. *Imagining early modern London: perceptions and portrayals of the city from Stow to Strype, 1598–1720*. Cambridge: Cambridge University Press.

Meyer, David R. 2006. *Networked machinists: High-technology industries in Antebellum America*. Baltimore: Johns Hopkins University Press.

Michie, Ranald. 2001. *The London Stock Exchange: a history*. Oxford: Oxford University Press.

Micklethwait, John, and Adrian Wooldridge. 2003. *The company: a short history of a revolutionary idea*. New York: Modern Library.

Middleton, John. 1798. *View of the agriculture of Middlesex with observations on the means of its improvement, and several essays on agriculture in general: drawn up for the consideration of the Board of Agriculture and Internal Improvement*. London: printed by B. Macmillan.

Mille, Adolphe Auguste. 1854. *Rapport sur le mode d'assainissement des villes en Angleterre et en Écosse présenté à M. Le Préfet de la Seine*. Paris: Vinchon.

Millward, Robert. 2005. *Private and public enterprise in Europe: energy, telecommunications and transport, 1830–1990*. New York: Cambridge University Press.

Minutes of evidence taken before the select committee on the supply of water to the metropolis. Parliamentary Papers (706). 1821.

Mirabeau, Honoré-Gabriel de Riquetti. 1786. *Sur les actions de la Compagnie des eaux de Paris*. 2nd ed. London.

Moeck-Schlömer, Cornelia. 1998. *Wasser für Hamburg: die Geschichte der Hamburger Feldbrunnen und Wasserkünste vom 15. bis zum 19. Jahrhundert*. Hamburg: Verein für Hamburgische Geschichte.

Mohajeri, Shahrooz. 2005. *100 Jahre Berliner Wasserversorgung und Abwasserentsorgung: 1840–1940, Blickwechsel.* Stuttgart: Steiner.

Moher, J. G. 1986. "John Torr Foulds (1742–1815): millwright and engineer." *Transactions of the Newcomen Society* 58: 59–73.

Mokyr, Joel. 2002. *The gifts of Athena: historical origins of the knowledge economy.* Princeton, N.J.: Princeton University Press.

——. 2009. *The enlightened economy: an economic history of Britain, 1700–1850.* New Haven, Conn.: Yale University Press.

Monconys, Balthasar de. 1665. *Iovrnal des voyages de Monsievr de Monconys où les sçauants trouueront vn nombre infini de nouueautez, en machines de mathematique, experiences physiques, raisonnemens de la belle philosophie, curiositez de chymie, & conuersations des illustres de ce siecle.* Lyon: Horace Boissat & George Remevs.

Morland, Samuel, and Joseph Morland. 1697. *Hydrostaticks: or, Instructions concerning water-works.* London: printed for John Lawrence.

Morris, Thomas. 1703. *The case of Thomas Morris, Esq.* London.

Mossoff, Adam. 2000. "Rethinking the development of patents: an intellectual history, 1550–1800." *Hastings Law Journal* 52: 1255–1322.

Mukhopadhyay, Asok Kumar. 1981. *Politics of water supply: the case of Victorian London.* Calcutta: World Press.

Mullett, Charles F. 1946. "Public baths and health in England, 16th–18th century." *Bulletin of the History of Medicine Supplement* 5: 1–85.

Murphy, Anne L. 2009. *The origins of English financial markets: investment and speculation before the South Sea Bubble.* Cambridge: Cambridge University Press.

——. 2013. "Financial markets: the limits of economic regulation in early modern England." In *Mercantilism reimagined: political economy in early modern Britain and its empire,* ed. P. J. Stern and C. Wennerlind, 263–82. Oxford: Oxford University Press.

——. 2014. "The financial revolution and its consequences." In *The Cambridge economic history of modern Britain,* vol. 1: *1700–1860,* ed. R. Floud, J. Humphries, and P. Johnson, 321–43. Cambridge: Cambridge University Press.

Murray, David. 1883. *The York Buildings Company: a chapter in Scotch history, read before the Institutes of Bankers and Chartered Accountants, Glasgow, 19th February, 1883.* Glasgow: James MacLehouse & Sons.

A new and complete dictionary of arts and sciences comprehending all the branches of useful knowledge. 1763. 2nd ed. London: printed for W. Owen, Society of Arts.

North, Douglass C., and Robert Paul Thomas. 1973. *The rise of the Western world: a new economic history.* Cambridge: Cambridge University Press.

North, Douglass C., and Barry R. Weingast. 1989. "Constitutions and commitment: the evolution of institutional governing public choice in seventeenth-century England." *Journal of Economic History* 49: 803–32.

Nugent, Thomas. 1749. *The grand tour: containing an exact description of most of the cities, towns, and remarkable places of Europe.* 4 vols. London: printed for S. Birt.

Nuvolari, Alessandro. 2004. "Collective invention during the British industrial revolution: the case of the Cornish pumping engine." *Cambridge Journal of Economics* 28: 347–63.

Observations upon the bill now depending, for supplying the cities of London and Westminster, and places adjacent, with water. 1725. London.

Overall, H. C., ed. 1878. *Analytical index, to the series of records known as the Remembrancia. Preserved among the archives of the city of London, A.D. 1579–1664*. London: E. J. Francis.

Owen, Edward. 1754. *Observations on the earths, rocks, stones and minerals for some miles about Bristol, and on the nature of the hot-well, and the virtues of its water*. London: printed and sold by W. Johnston, at the Golden-Ball in St. Paul's Church-Yard.

Park, John James. 1814. *The topography and natural history of Hampstead, in the county of Middlesex*. London: printed for White Cochrane.

Parliamentary Papers. parlipapers.chadwyck.co.uk.

Pearson, Robin. 2004. *Insuring the industrial revolution: fire insurance in Great Britain, 1700–1850*. Aldershot: Ashgate.

Pearson, Robin, and David Richardson. 2001. "Business networking in the industrial revolution." *Economic History Review* 54: 657–79.

Peck, Linda Levy. 2005. *Consuming splendor: society and culture in seventeenth-century England*. Cambridge: Cambridge University Press.

Penny Cyclopaedia of the Society for the Diffusion of Useful Knowledge. 1833. 27 vols. London: Charles Knight.

Peppercorne, Frederick S. 1840. *Supply of water to the Metropolis. A brief description of the various plans that have been proposed for supplying the metropolis with pure water*. London: John Weale.

Percival, Thomas. 1769. *Experiments and observations on water particularly on the hard pump water of Manchester*. London: printed for J. Johnson in Pater-Noster-Row.

Périer, Jacques-Constantin. 1777. *Distribution d'eau de la Seine dans tous les quartiers et dans toutes les maisons de Paris*. Paris: La veuve Ballard.

Pettigrew, William A. 2011. "Regulatory inertia and national economic growth: an Africa trade case study, 1690–1714." In *Regulating the British economy, 1660–1850*, ed. P. Gauci, 25–40. Farnham, Surrey: Ashgate.

Petty, William. 1683. *Another essay in political arithmetick, concerning the growth of the city of London with the measures, periods, causes, and consequences thereof: 1682*. London: printed by H.H. for Mark Pardoe.

Pollard, Sidney. 1965. *The genesis of modern management: a study of the industrial revolution in Great Britain*. Cambridge, Mass.: Harvard University Press.

Porter, Roy. 1995. "The eighteenth century." In *The Western medical tradition: 800 BC to AD 1800*, ed. L. I. Conrad, 371–476. Cambridge: Cambridge University Press.

Price, Richard. 1773. *Observations on reversionary payments*. 3rd ed. London: printed for T. Cadell.

———. 1780. *An essay on the population of England from the Revolution to the present time. With an appendix*. 2nd ed. London: printed for T. Cadell.

Price, Richard, and William Morgan. 1803. *Observations on reversionary payments on schemes for providing annuities for widows, and for persons in old age: on the method of calculating the values of assurances on lives*. 6th ed. 2 vols. London: T. Cadell and W. Davies.

Price, Robin. 1981. "Hydropathy in England 1840–70." *Medical History* 25: 269–80.

Pringle, John. 1752. *Observations on the diseases of the army, in camp and garrison. In three parts*. London: printed for A. Millar, and D. Wilson.

Prospectus de la fourniture et distribution des eaux de la Seine, à Paris, par les machines à feu. 1781. Paris: La veuve Ballard.

Pumfrey, Stephen. 1995. "Who did the work? Experimental philosophers and public demonstrators in Augustan England." *British Journal for the History of Science* 28: 131–56.

Radkau, Joachim. 1994. "Zum ewigen Wachstum verdammt? Jugend und Alter großer technischer Systeme." In *Technik ohne Grenzen,* ed. I. Braun and J. Bernward, 50–106. Frankfurt am Main: Suhrkamp.

The Raymond and Beverly Sackler Archive Resource. Lowthorp; John (c 1659–1724). 2013. Royal Society. royalsociety.org/library/collections/biographical-records.

Reddaway, T. F. 1938. "London bridge waterworks after the Great Fire of 1666." *Transactions of the Newcomen Society* 19: 217–19.

———. 1940. *The rebuilding of London after the great fire.* London: Jonathan Cape.

Reden, Friedrich. 1851. *Allgemeine vergleichende finanz-statistik.* Darmstadt: G. Jonghaus.

Register van Holland en Westvriesland. Van den jaare 1586 en 1587. 1587. 's-Gravenhage.

Relton, Francis Boyer. 1893. *An account of the fire insurance companies associations institutions projects and schemes established and projected in Great Britain and Ireland during the 17th and 18th centuries including the Sune Fire Office.* London: Swan Sonnenschein.

Report of the commissioners appointed by his majesty to inquire into the state of the supply of water in the metropolis. April 21, 1828.

Reynolds, Susan. 2010. *Before eminent domain: toward a history of expropriation of land for the common good.* Studies in legal history. Chapel Hill: University of North Carolina Press.

Reynolds, Terry S. 1983. *Stronger than a hundred men: a history of the vertical water wheel.* Baltimore: Johns Hopkins University Press.

Richards, H. C., William Henry Christopher Payne, and John Philpott Henry Soper. 1899. *London water supply Being a compendium of the history, law, & transactions relating to the metropolitan water companies. Second edition.* London: P. S. King & Son Orchard House Westminster.

Richmond, Lesley, and Alison Turton, eds. 1990. *The Brewing Industry: A Guide to Historical Records.* Manchester: Manchester University Press.

Riley, Henry T. 1868. *Memorials of London and London life in the XIIIth, XIV, and XVth centuries: being a series of extracts, local, social, and political from the early archives of the city of London. A.D. 1276–1419.* London: Longmans.

Roberts, M. J. D. 1988. "Public and private in early nineteenth-century London: The Vagrant Act of 1822 and its enforcement." *Social History* 13: 273–94.

Roberts, Walter. 1641a. *A proposition for the serving and supplying of London and Westminster and other places adjoyning with a sufficient quantity of good and cleare spring water, to be brought from Huddesdon in Hartfordshire in a close aqueduct of bricke, stone, lead, or timber.*

———. 1641b. *Sir VValter Roberts his Ansvver to Mr. Fords book, entituled, A designe for bringing a navigable river, from Rickmansworth in Hartfordshire to St. Giles in the Fields.* London: printed by R.H.

Robertson, H. M. 1934. "Sir Bevis Bulmer: a large-scale speculator of Elizabethan and Jacobean times." *Journal of Economic and Business History* 4: 99–120.

Robins, Frederick William. 1946. *The story of water supply*. Oxford: Oxford University Press.

Robison, John. 1797. "Water-works." In *Encyclopædia Britannica or, a dictionary of arts, sciences, and miscellaneous literature*. Edinburgh: printed for A. Bell and C. Macfarquhar.

Rolle, Samuel. 1667. *Shlohavot, or, The burning of London in the year 1666 commemorated and improved in a CX discourses, meditations, and contemplations*. London: printed by R.I. for Nathaniel Ranew and Jonathan Robinson.

Rosenberg, Nathan, and L. E. Birdzell. 1986. *How the West grew rich: the economic transformation of the industrial world*. New York: Basic Books.

Roseveare, Henry. 1991. *The financial revolution, 1660–1760*. London: Longman.

Royal Commission on Historical Manuscripts. 1858. *Calendar of State Papers, Domestic Series, of the reign of James I, 1619–1623, preserved in the State Paper Department of her Majesty's Public Record Office*. Vol. 3: *1619–1623*. Ed. M. A. E. Green. London: Longman, Brown, Green, Longmans and Roberts.

——. 1872. *Third report of the Royal commission on historical manuscripts*. London: H.M.S.O. by Eyre and Spottiswoode.

——. 1874. *Fourth report of the Royal commission on historical manuscripts*. London: H.M.S.O. by Eyre and Spottiswoode.

——. 1877. *Sixth report of the Royal commission on historical manuscripts*. London: H.M.S.O.

——. 1894a. *Calendar of State Papers, Domestic Series, of the reign of Charles II, 1668–1669, preserved in the State Paper Department of Her Majesty's Public Record Office*. Vol. 9: *Oct 1668–Dec 1669*. Ed. M. A. E. Green. London: Longman, Brown, Green, Longmans and Roberts.

——. 1894b. *The manuscripts of His Grace the Duke of Portland, preserved at Welbeck Abbey*. Vol. 3: *Report*. London: printed for H.M.S.O. by Eyre and Spottiswoode etc.

——. 1894c. *The manuscripts of the House of Lords, 1692–1693*. 14th report, appendix, part 6. London: printed for H.M.S.O. by Eyre and Spottiswoode.

——. 1897. *The manuscripts of His Grace the Duke of Portland, preserved at Welbeck Abbey*. Vol. 4. London: printed for H.M.S.O. by Eyre and Spottiswoode etc.

——. 1910. *Calendar of the Cecil Papers in Hatfield House*. Vol. 12: *1602–1603*. London: H.M.S.O.

——. 1933a. *Calendar of the Cecil Papers in Hatfield House*. Vol. 16: *1604*. London: H.M.S.O.

——. 1933b. *Calendar of the manuscripts of the most Hon. the Marquis of Salisbury*. Vol. 16: *1604*. Ed. M. S. Guiseppi. London: H.M.S.O.

——. 1938. *Calendar of the manuscripts of the Most Hon. the Marquis of Salisbury, preserved at Hatfield House, Hertfordshire*. Vol. 17: *1605*. Ed. M. S. Guiseppi. London: H.M.S.O.

Rudden, Bernard. 1985. *The New River: a legal history*. Oxford: Clarendon Press.

Salmon, Thomas. 1743. *The history and present state of the British islands describing their respective situations, buildings, customs*. London: printed for J. Robinson.

Sanderson, Roberto. 1732. *Foedera, Conventiones, Literae, Et Cujuscunque generis acta publica, inter reges Angliae*. London: J. Tonson.

Saussure, Cesar de, and Madame Van Muyden. 1902. *A foreign view of England in the reigns of George I and George II: the letters of Monsieur Cesar de Saussure to his family*. London: J. Murray.

Schnitter, Niklaus. 1992. *Die Geschichte des Wasserbaus in der Schweiz.* Oberbözberg: Olynthus.

Schwarz, L. D. 1992. *London in the age of industrialisation: entrepreneurs, labour force, and living conditions, 1700–1850.* Cambridge: Cambridge University Press.

———. 2000. "London, 1700–1840." In *The Cambridge urban history of Britain*, vol. 2: *1540–1840*, ed. P. Clark, 315–46. Cambridge: Cambridge University Press.

Scott, William Robert. 1910. *The constitution and finance of English, Scottish and Irish joint-stock companies to 1720.* 3 vols. Cambridge: Cambridge University Press.

Seymour, Robert, and John Stow. 1733. *A survey of the cities of London and Westminster, borough of Southwark, and parts adjacent.* London: T. Read.

Sharpe, Reginald R. 1894. *London and the kingdom; a history derived mainly from the archives at Guildhall in the custody of the corporation of the city of London.* Vol. 2. London: Longmans Green.

Sheard, Sally. 2000. "Profit is a dirty word: the development of public baths and wash-houses in Britain 1847–1915." *Social History of Medicine* 13: 63–86.

Sheppard, F. H. W. 1998. *London: a history.* New York: Oxford University Press.

"Sir Hugh Middleton's Scheme for supplying the City of London with good and whole-some Water." 1751. *Universal Magazine* 8: 309–12.

Sisley, Richard. 1899. *The London water supply: a retrospect and a survey.* London.

Skempton, A. W. 2002. "Wright, Edward." In *A biographical dictionary of civil engineers in Great Britain and Ireland*, vol. 1: *1500–1830*, ed. A. W. Skempton. London: T. Telford.

Slack, Paul. 1986. "Metropolitan government in crisis: the response to plague." In *London 1500–1700: the making of the metropolis*, ed. A. L. Beier and R. Finlay, 60–81. London: Longman.

———. 2014. *The invention of improvement: information and material progress in seventeenth-century England.* Oxford: Oxford University Press.

Smart, John. 1742. *A short account of the several wards, precincts, parishes, &c. in London.* London.

Smeaton, John. 1812. *Reports of the late John Smeaton, F.R.S made on various occasions in the course of his employment as a civil engineer.* London: printed for Longman.

Smith, Norman Alfred Fisher. 1976. *Man and water: a history of hydro-technology.* London: P. Davies.

Smith, Virginia. 2007. *Clean: a history of personal hygiene and purity.* Oxford: Oxford University Press.

Spinage, C. A. 2003. *Cattle plague: a history.* New York: Kluwer Academic / Plenum Publishers.

Squatriti, Paolo, ed. 2000. *Working with water in medieval Europe: technology and resource-use.* Leiden: Brill.

Stern, Philip J. 2013. "Companies: monopoly, sovereignty, and the East Indies." In *Mercantilism reimagined: political economy in early modern Britain and its empire*, ed. P. J. Stern and C. Wennerlind, 177–96. Oxford: Oxford University Press.

Stewart, Larry. 1992. *The rise of public science: rhetoric, technology, and natural philosophy in Newtonian Britain.* New York: Cambridge University Press.

Stow, John, and Charles Lethbridge Kingsford. 1603 [1908]. *A Survey of London, by John Stow, reprinted, from the text of 1603.* Oxford: Clarendon Press.

Stow, John, and Anthony Munday. 1633. *The survey of London containing the original, increase, modern estate and government of that city, methodically set down*. London: printed for Nicholas Bourn.

Stow, John, and John Strype. 1720. *A survey of the cities of London and Westminster: containing the original, antiquity, increase, modern estate and government of those cities*. 2 vols. London: printed for A. Churchill.

Summerson, John, and Howard Colvin. 2003. *Georgian London*. New Haven, Conn.: Yale University Press.

Summerton, Jane, ed. 1994. *Changing large technical systems*. Boulder, Colo.: Westview Press.

Sussman, Nathan, and Yishay Yafeh. 2006. "Institutional reforms, financial development and sovereign debt: Britain 1690–1790." *Journal of Economic History* 66: 906–35.

Switzer, Stephen. 1729. *An introduction to a general system of hydrostaticks and hydraulicks, philosophical and practical*. 2 vols. London: printed for T. Astley.

Szostak, Rick. 1991. *The role of transportation in the industrial revolution: a comparison of England and France*. Montreal: McGill-Queen's University Press.

Tabberner, John Loude. 1847. *The past, the present, and the probable future supply of water to London with observations in respect of improved sanitary measures*. London: Longman.

Taylor, Vanessa, and Frank Trentmann. 2011. "Liquid politics: water and the politics of everyday life in the modern city." *Past & Present* 211: 199–241.

Thirsk, Joan. 1978. *Economic policy and projects: the development of a consumer society in early modern England*. Oxford: Clarendon Press.

Thomas, J. H. 1979. "Thomas Neale, a seventeenth-century projector." Ph.D. diss., University of Southampton.

Thomas, Keith. 1997. "Health and morality in early modern England." In *Morality and health*, ed. A. M. Brandt and P. Rozin, 15–34. New York: Routledge.

Thomson, Richard. 1827. *Chronicles of London Bridge*. London: Smith Elder.

Thornbury, Walter, and Edward Walford. 1878. *Old and new London*. London: Cassels.

Thorsheim, Peter. 2006. *Inventing pollution: coal, smoke, and culture in Britain since 1800*. Athens: Ohio University Press.

Thrush, A. D., and John P. Ferris. 2010. *The House of Commons, 1604–1629*. Cambridge: published for the History of Parliament Trust by Cambridge University Press.

Thumb, Tom. 1746. *The travels of Tom Thumb over England and Wales; containing descriptions of whatever is most remarkable in the several counties*. London: printed for R. Amey.

Tijdschrift voor geschiedenis, oudheden en statistiek van Utrecht: met naamlijst der geborenen, ondertrouwden en overledenen binnen Utrecht en voorsteden. 1841. Vol. 7. Utrecht: N. van der Monde.

Tomory, Leslie. 2012. *Progressive enlightenment: the origins of the gaslight industry, 1780–1820*. Cambridge, Mass.: MIT Press.

Trew, Alexander. 2010. "Infrastructure finance and industrial takeoff in the United Kingdom." *Journal of Money, Credit and Banking* 42: 985–1010.

A true copy of several affidavits and other proofs of the largeness and richness of the mines, late of Sir Carbery Pryse the original whereof are fil'd in the High Court of Chancery. 1698. London: printed by Freeman Collins.

Trusler, John. 1786. *The London adviser and guide containing every instruction and information useful and necessary to persons living in London, and coming to reside there.* London: printed for the author.

Tryon, Thomas. 1683. *The way to health, long life and happiness, or, A discourse of temperance and the particular nature of all things requisit for the life of man, as all sorts of meats, drinks, air, exercise.* London: Sowle.

Turner, G. L'E. 2004. "Baker, Henry (1698–1774)." In *Oxford dictionary of national biography*, ed. H. C. G. Matthew, B. Harrison, and L. Goldman. Oxford: Oxford University Press.

Turner, Henry S. 2013. "Corporations: humanism and Elizabethan political economy." In *Mercantilism reimagined: political economy in early modern Britain and its empire*, ed. P. J. Stern and C. Wennerlind, 154–77. Oxford: Oxford University Press.

Tynan, Nicola. 2002. "London's private water supply, 1582–1902." In *Reinventing water and wastewater systems: global lessons for improving water management*, ed. P. Seidenstat, D. Haarmeyer, and S. Hakim, 341–60. New York: John Wiley and Sons.

The Universal British directory of trade and commerce Comprehending lists of the inhabitants of London, Westminster, and borough of Southwark; and of all the cities, towns, and principal villages, in England and Wales. 1790. London: printed for the Patentees, and sold by C. Stalker.

Vachette, J., and P. Vachette. 1791 [1797]. *Précis historique sur l'établissement des pompes à feu des sieurs Perier frères à Paris, leur manutention, régie, agiotage et les autres abuts de l'administration.* Paris.

van Lieshout, Carry. 2012. "London's changing waterscapes—the management of water in eighteenth-century." Ph.D. diss., University of London.

Vernon, Thomas. 1726. *Cases argued and adjudged in the High Court of Chancery Published from the Manuscripts of Thomas Vernon, Late of the Middle Temple, Esq; By order of the High Court of Chancery.* Vol. 1. Dublin: printed by J. Watts.

Vigarello, Georges, and Jean Birrell. 1988. *Concepts of cleanliness: changing attitudes in France since the Middle Ages.* Cambridge: Cambridge University Press.

Volkmann, Johann Jacob. 1782. *Neueste Reisen durch England.* Leipzig: Fritsch.

Voltaire, Beaumarchais Pierre Augustin Caron de. 1784. *Oeuvres completes de Voltaire.* Ed. J.-A.-N. d. C. Condorcet and J. J. M. Decroix. 70 vols. Kehl: De L'Imprimerie de la Société Littéraire-Typographique.

Von Tunzelmann, G. N. 1978. *Steam power and british industrialization to 1860.* Oxford: Clarendon Press.

Voth, Hans-Joachim. 2000. *Time and work in England, 1750–1830.* Oxford: Clarendon Press.

Waller, William. 1698. *An essay on the value of the mines, late of Sir Carbery Price.* London.

Wallis, P. J. 1976. "Wright, Edward." In *Complete dictionary of scientific biography.* Detroit: Charles Scribner's Sons.

Walton, Phillippa. 1738. *The case of the petitioners, Phillippa and John Walton, gun-powder-makers; owners, and occupiers of the gun-powder-mills, situated at Waltham-Abbey.*

Ward, Frederick Oldfield. 1850. "Metropolitan water supply." *London Quarterly Review* 87: 253–72.

———. 1880. "Address of Mr. F. O. Ward, General Congress of hygiene at Brussels." In *Report of the fourth congress of the sanitary institute of Great Britain, held at Exeter September 1880*, 267–71. London: Office of the Institute.

Ward, J. R. 1974. *The finance of canal building in eighteenth-century England.* Oxford: Oxford University Press.

Ward, Robert. 2003. *London's New River.* London: Historical Publications.

Warner, Sam Bass. 1968. *The private city: Philadelphia in three periods of its growth.* Philadelphia: University of Pennsylvania Press.

Water. 1819. In *Cyclopædia; or, Universal dictionary of arts, sciences, and literature.*, ed. A. Rees. London: Longman.

Weale, James. 1820. *Water monopoly. The case of the water companies stated and examined, or, The calm address dissected with remarks thereon, calculated to settle opinions on the subject of the additional water rates.* London: printed by C. Richards.

Weinstein, Rosemary. 1991. "New urban demands in early modern London." *Medical History Supplement* 11: 29–40.

Wengenroth, Ulrich. 2003. "Science, technology, and industry." In *From natural philosophy to the sciences: writing the history of nineteenth-century science,* ed. D. Cahan, 221–53. Chicago: University of Chicago Press.

White, Charles. 1773. *A treatise on the management of pregnant and lying-in women, and the means of curing, but more especially of preventing the principal disorders to which they are liable.* London: printed for Edward and Charles Dilly.

Wicksteed, Thomas. 1835. *Observations on the past and present supply of water to the Metropolis.* London: Richard Taylor.

Wilkinson, Robert, and Charles Lancelot Shadwell. 1819. *Londina illustrata: graphic and historic memorials of monasteries, churches, chapels, schools, charitable foundations, palaces, halls, courts, processions, places of early amusement and modern and present theaters.* 2 vols. London: published by Robt. Wilkinson.

Willan, Thomas Stuart. 1936. *River navigation in England, 1600–1750.* London: H. Milford.

Williamson, F. 1936. "Sorocold of Derby, a pioneer of water supply." *Journal of the Derbyshire Archaeological and Natural History Society* 10: 43–93.

Williamson, F., and W. B. Crump. 1945. "Sorocold's Waterworks at Leeds." *Thoresby Miscellany* 11: 166–82.

Williamson, Jeffrey G. 1982. "The structure of pay in Britain, 1710–1911." *Research in Economic History* 7: 1–54.

Wilson, Adrian. 1995. *The making of man-midwifery: childbirth in England, 1660–1770.* London: UCL Press.

Wilson, J. F. 1995. *British business history, 1720–1994.* Manchester: Manchester University Press.

Winfield, Rif. 2007. *British warships in the age of sail, 1714–1792: design, construction, careers and fates.* 2nd rev. ed. Barnsley: Seaforth Publishing.

———. 2009. *British warships in the age of sail, 1603–1714: design, construction, careers and fates.* Barnsley: Seaforth Publishing / MBI Publishing.

Winsor, Frederick Albert. 1807. *To be sanctioned by act of parliament. A National Light and Heat Company, for providing our streets and houses with hydrocarbonic gas-lights, on similar principles, as they are now supplied with water.* New ed. London: printed by Watts & Bridgewater.

Woodley, Roger. 2004. "Mylne, Robert (1733–1811), architect and engineer." In *Oxford dictionary of national biography,* ed. H. C. G. Matthew, B. Harrison, and L. Goldman. Oxford: Oxford University Press.

Worboys, Michael. 2007. "Was there a bacteriological revolution in late nineteenth-century medicine?" *Studies in History and Philosophy of Science Part C: Studies in History and Philosophy of Biological and Biomedical Sciences* 38: 20–42.

Wren, Christopher. 1753. "Thoughts of Sir Christopher Wren on the distribution of New River Water." *Gentleman's Magazine, and Historical Chronicle* 23: 114–16.

Wrigley, E. A. 1985. "Urban growth and agricultural change: England and the Continent in the early modern Period." *Journal of Interdisciplinary History* 15: 683–728.

Żelichowski, Ryszard. 2004. "Lindley, William (1808–1900)." In *Oxford dictionary of national biography*, ed. H. C. G. Matthew, B. Harrison and L. Goldman. Oxford: Oxford University Press.